The Wind Blew Me There

3/22/19

Forrest and Rachael,

Enjoy this trip with me!

Barney

The Wind Blew Me There

Memories of a Ship's Surgeon Aboard Barquentine *Verona*

Barnett L. Cline, MD

ISBN-13: 9781979204880
ISBN-10: 1979204888
Library of Congress Control Number: 2017916907
CreateSpace Independent Publishing Platform
North Charleston, South Carolina

Table of Contents

The Barquentine *Verona*'s Ports of Call between Panama and Lisbon, Portugal

Galápagos Islands
- San Cristóbal (Wreck Bay)
- Santa Cruz (Academy Bay)
- Española (Gardner Bay)
- Floreana (Post Office Bay)
- Genovesa (Darwin Bay)
- Santiago (Sullivan Bay, James Bay)
- Isabela (Elizabeth Bay)

Pitcairn Islands (Bounty Bay)

French Polynesia
- Marquesas Islands
- Nuku Hiva
- Tahiti (Papeete Harbor)
- Huahine (Fare, Hapu)
- Tahaa (Faaaha, Teanorea Bay, Patio)
- Bora-Bora

Cook Islands (Palmerston Atoll)

Samoa (Apia)

Fiji
- Suva
- Mbengha Island

Vanuatu (New Hebrides)
- Efate Island (Vila Harbor)
- Tanna Island
  - Lenakel Bay
  - Black Beach
- Malekula Island
  - Tesman Bay
  - Esprigle Bay
- Ambrym Island (Redds Anchor)
- Espíritu Santo Island (Luganville)

Papua New Guinea
    Samarai Island
    Esa'ala
    Trobriand Islands
        Iwa Island
        Kiriwina (Kaibola Village)
        Losuia
    Madang
Taiwan (Formosa)
    Kaohsiung
Hong Kong (Kowloon Harbor)
Thailand (Bangkok)
Malaysia (Penang Island, Georgetown)
Andaman Islands
    Port Blair
    North Sentinel Island
India (Chennai; was Madras)
Sri Lanka (was Ceylon) (Trincomalee)
Maldive Islands
    Malé
    Horsburgh Atoll
Yemen (Aden)
Saudi Arabia (Umm Lajj)
Egypt
    Suez
    Port Said
Lebanon (Beirut)
Greece
    Rhodes
    Lindos
    Astypalaia
    Santorini

Malta (Valletta)
Spain (Cádiz)
Portugal (Lisbon)

List of Maps

List of Photographs

Preface

A youth with a heartbeat cannot circle most of the globe on a sailboat for a year without being profoundly altered. For decades, I did not adequately appreciate the uniqueness of that experience, but now, aided by some fifty years of perspective, I more clearly see how it enriched my life. The opportunity to join the *Verona*'s crew came unexpectedly, when in 1965, as a young physician, I had just completed my military service obligation.

Because memory suffers after such an interval, my re-exploration of that voyage has been facilitated by my daily log, the many photographs I took, a highly selective collection of old letters, and photography notes. These source materials survived and not only assisted me immensely in recalling remote experiences, but they also permitted detail and accuracy otherwise not possible.

Parts of the globe I came to know have evolved during the intervening years, as have I; some places I have revisited. Once experienced, these distant lands have continued to intrigue me, and for many, I have tried to maintain current knowledge. In addition to sharing memorable experiences and ports of call, my story introduces the *Verona*'s unique Captain Christopher Sheldon, some of my crewmates, and others I met along the way, describing how their lives were similarly touched by the *Verona*.

For years now, I have promised this memoir to family and friends as a means to leave behind traces of a year most of us can only dream of. What began as an effort to capture high adventures from a sea voyage, writing *The Wind Blew Me There,* led to reflection and introspection that grew into this longer, very personal journey.

Most of the book's chapters focus upon a single place, person, or event, not necessarily in chronological order. The three "Peripatetic Dreaming" chapters reflect travel in different geographic regions and are ordered chronologically. They contain verbatim quotes from my log book, other notes, and expanded descriptions of some of the *Verona*'s destinations.

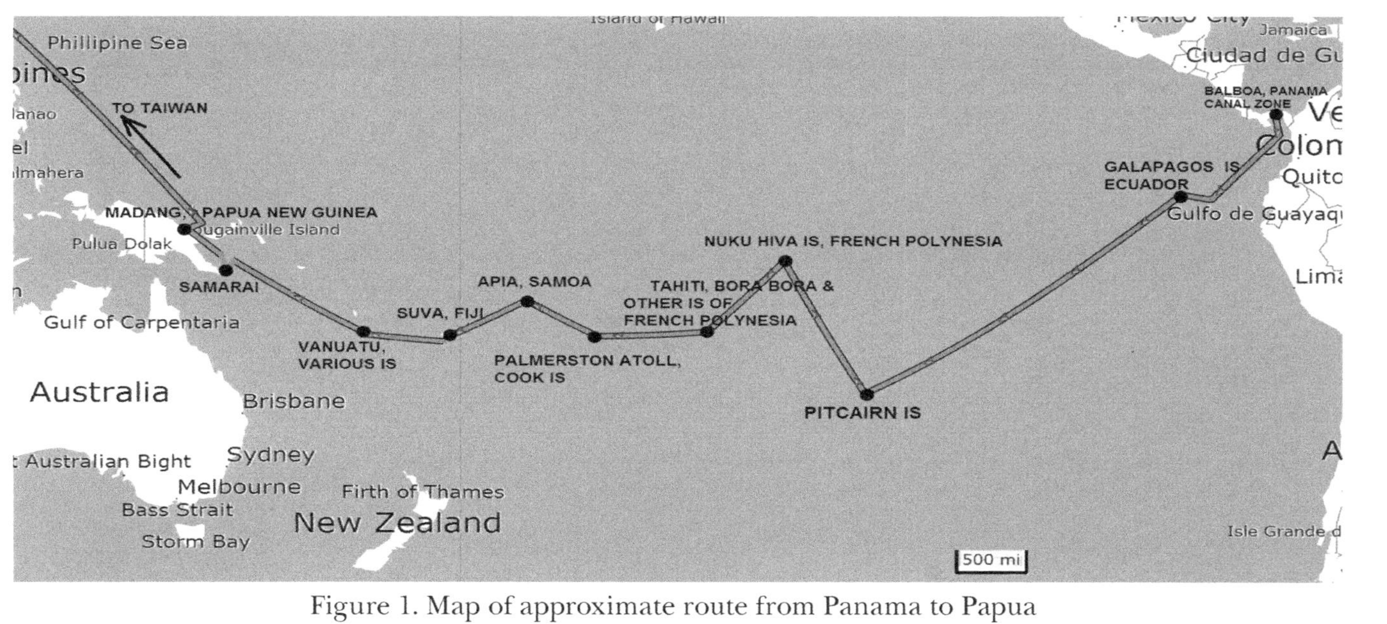

Figure 1. Map of approximate route from Panama to Papua New Guinea. Prepared by Gerald Cline.

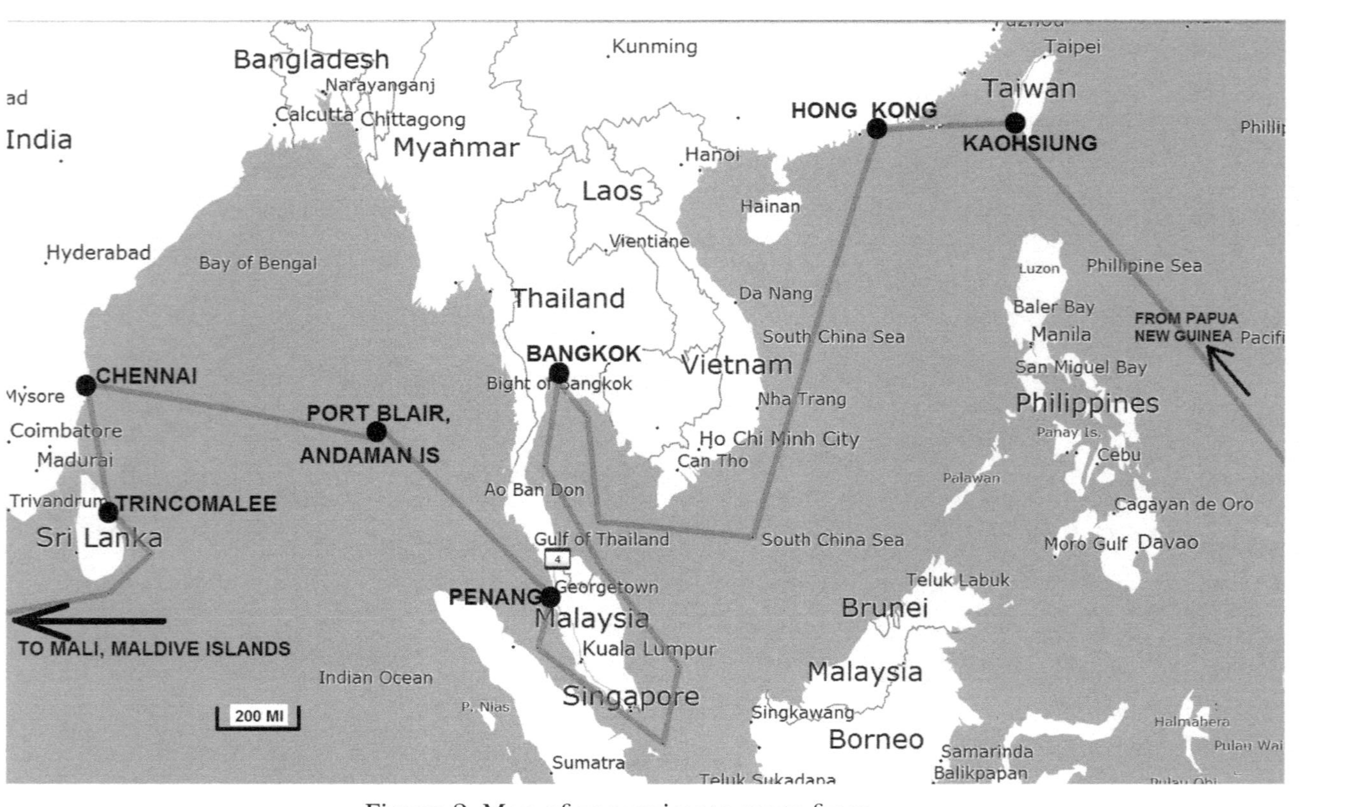

Figure 2. Map of approximate route from
New Guinea to Sri Lanka. Prepared by Gerald Cline.

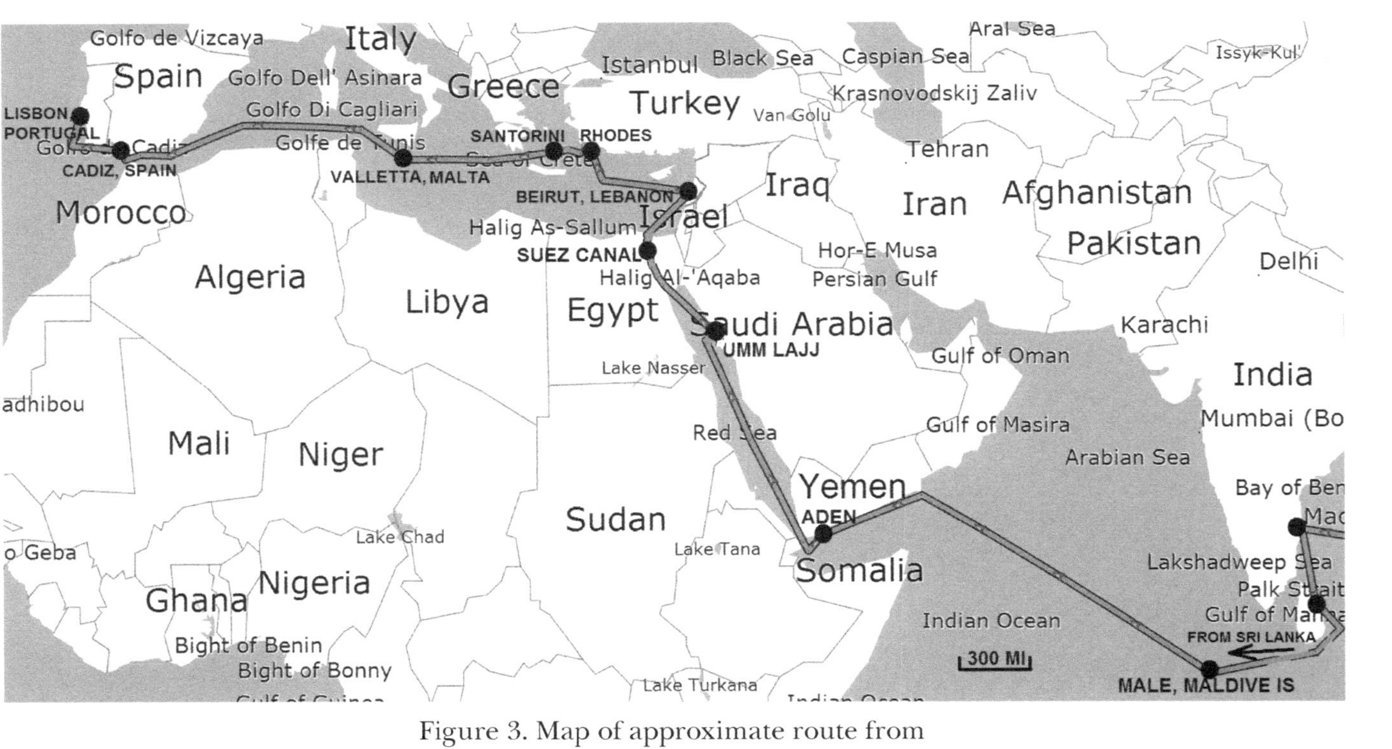

Figure 3. Map of approximate route from
Sri Lanka to Portugal Prepared by Gerald Cline.

One

Captain Christopher Sheldon and the Barquentine *Verona*

When I first read his obituary in the October 29, 2002 New York Times I shuddered, recalling when I met Chris in Bogota, Columbia. I remembered the thrill and chill that ran through me when he invited me to join the *Verona's* crew as the doctor on a year-long sailing adventure. My thoughts focused on this proud man, graced with physical and intellectual power, who having been tested by terrifying adversity struggled to the end of his life to overcome ill fortune and internal demons. But on June 26, 1965 as the *Verona* sailed from Panama into the vast Pacific Ocean I was blissfully unaware that dark clouds followed our skipper, my confidence in him unqualified.

Key portions of his obituary are shown below:

"***C.B. Sheldon, 76, Skipper in a Noted Sinking***
Christopher B. Sheldon, whose 92-foot twin-masted sailing ship Albatross sank in a freak storm in 1961, a disaster that inspired the 1996 movie "White Squall," died on Oct. 5 in Stamford, Conn. He was 76 and lived in Norwalk, Conn.

Mr. Sheldon was shepherding high-school students on the ship, used as his floating classroom, when a horrific squall welled up and abruptly sank it. Jeff Bridges portrays Mr. Sheldon in the movie about the sinking, in which his wife, Alice, and five others were lost.

About 8:30 a.m. on May 2, 1961, the Albatross was gliding through a slight mist in calm seas 180 miles west of Key West on the way to Nassau. Suddenly, a single bolt of lightning flashed across the sky, and a blast of wind smacked the ship.

"It was as if a giant hand took hold of us," Mr. Sheldon said in a 1996 interview with People magazine. "In 15 seconds the Albatross was on its side. In 60 seconds it filled with water. And then it was gone—the ocean was calm."

Four students, the ship's cook and Mrs. Sheldon had vanished. Mr. Sheldon, 11 students and an English and math teacher scrambled into lifeboats. They were rescued a day later by a passing Dutch freighter.

When the storm hit, Mr. Sheldon did not even have time to begin to take down the sail. It all became such a blur that he could not remember flying home.

"He did the best he could," Dick Langford, the teacher who survived, said after the sinking. "He was just a bad-luck skipper."

In May 1959, he married Alice N. Strahan, a doctor who had served as medical officer on the Yankee. Together they founded the Ocean Academy as a floating prep school in 1959. He held that the sea was "a great molder of character." Students paid $3,250 for an academic year of study, including Spanish and celestial navigation, taught by Mr. Sheldon.

When the white squall hit the Albatross, the six people who were below deck could not escape.

After the survivors dispersed, they did not see each other until some gathered to help make the movie, directed by Ridley Scott.

The screenplay was based on a 1963 book about the incident, "The Last Voyage of the Albatross," written by Chuck Gieg, one of the students. Other survivors included William P. Bunting, son of Mary I. Bunting, the president of Radcliffe College.

After the accident, Mrs. Bunting told Sargent Shriver, the first director of the Peace Corps, about Mr. Sheldon, and he offered him a job. He became director of operations for the Peace Corps in Colombia, a distraction he welcomed.

In addition to Ms. Ramsey, he is survived by a brother, John, of South Bristol, Me.

In 1965, he again developed sea fever and bought a 130-foot ship, the Verona, for use as a floating school. On his second voyage in the ship, the ship caught fire near the west coast of Central Africa. The fire destroyed the vessel, but all aboard escaped.[1]

This odd obituary inextricably connects Christopher "Chris" Sheldon with the "white squall" that in large measure defined his life. It also says that Chris accepted a position in Colombia as director of the Peace Corps in that country and that he welcomed this challenge as a distraction from his tragic loss of *Albatross* in 1961. Chris's obituary also makes prominent mention of the 1996 movie *White Squall*, starring Jeff Bridges as Sheldon, but fails to convey that the movie, with its Hollywood scripting, drastically altered reality.

---

1. Douglas Martin, "C.B. Sheldon, 76, Skipper in a Noted Sinking," *New York Times*, October 29, 2002, http://www.nytimes.com/2002/10/29/us/cb-sheldon-76-skipper-in-a-noted-sinking.html.

One *Albatross* survivor, Richard E. Langford, the English teacher on its final voyage, in 2001 published *White Squall: The Last Voyage of Albatross.* The prologue to Langford's book begins: "Back in 1960 I answered an ad in *Yachting* Magazine, placed by The Ocean Academy Ltd. owned and operated by Christopher B. Sheldon, Ph.D. and N. Alice Sheldon, M.D. They wanted a teacher of English for a nine-month voyage on the school ship *Albatross,* a square-rigged brig to be crewed by teen-aged students."[2] Later in the prologue, Langford states that he wrote most of the book during the mid-1960s and then put it aside. His prologue continues: "The film *White Squall* engendered fresh interest in the *Albatross* voyage, even though the film was more Hollywood parody than fact. Readers of this volume will acquire a more realistic understanding of the people and events involved."[3] And on the back cover of the book are the words, "You have seen the movie *White Squall,* now it is time to read the true account of that final, fatal voyage of the school ship *Albatross.*"[4]

Hoping to learn more about Chris from it, I pounced on Langford's book when I first learned of it several years ago and even succeeded in chatting with him by telephone at his home in Florida. The movie disappointed me, but Langford's small book was moving and full of useful insights into the circumstances of the *Albatross*'s sinking in 1961 and into the nature and character of Chris Sheldon. Perhaps Langford's book would reveal clues to the forces that ultimately led to the crisis the *Verona*'s crew faced years later in the Cook Islands and to the fiery fate of the *Verona*

---

2. Richard E. Langford, *White Squall: The Last Voyage of Albatross,* ed. Jerry Renninger (Harrisburg, PA: Bristol Fashion Publications, 2001), 9.

3. Ibid., 10.

4. Ibid., back cover.

> on its second voyage, when it burned and sank off the coast of Equatorial Guinea in late 1966.

The book's initial ninety pages offered a good account of life aboard a school ship, relating some of the problems and frustrations experienced by students and teachers, but not until page 91 did I encounter hints of concern about the safety of the vessel and its operation. Langford described the frightening experience of being awakened in the Galápagos Islands by the *Albatross*'s keel striking bottom several times. In daylight the next day, he was able to joke with a friend about creating a new drink called "*Albatross* on the rocks," but the captain was not amused.[5] Langford then went on to write, "this wasn't the only time we escaped possible serious injury or damage," describing a longboat that fell as it was being lowered over the side, but without injury.[6] He also noted that the *Albatross,* anchored in heavy swells off a long lava reef, had lost an anchor when a chain link broke, but disaster was averted. Langford cited this and other similar episodes as additional evidence for Joseph Conrad's contention that we are all "eternally menaced, most of all by ourselves."[7] Langford concludes, "a moment's carelessness or neglect, or a smug sense of wellbeing can often lead to disaster."[8]

Langford's final chapter describes the sinking of the school ship *Albatross.* In stark contrast to the description of the sinking in the obituary, he describes "a long, high, black squall line to the north" during his watch, which ended at 4 a.m.[9] Awake at 7:30 a.m., he could hear rough water on the steel hull and had trouble climbing out of his starboard bunk because the *Albatross* was

5. Ibid., 91.
6. Ibid., 91-92.
7. Ibid., 92.
8. Ibid., 92.
9. Ibid., 97.

heeled sharply to starboard. "At breakfast, the main cabin table tilted sharply in its gimbals as *Albatross* rolled through roughening waters. The lightning was closer, and we could hear loud booms of thunder very near."[10] As rain spattered on Chris Sheldon's nose through the skylight, he (Sheldon) said, "we might get some fun out of that in a little while. Good to hear the wind again."[11] Not long afterwards the *Albatross* capsized. I found myself dumfounded to read such disparate accounts of the fatal "white squall" as those of two survivors, Richard Langford and Christopher Sheldon.

Bogotá is where this appealing, enigmatic, and impenetrable man, who would become my captain, entered my life. But the story started in Bolivia, a landlocked country in the Andes, which seems an unlikely place for the genesis of a year at sea. It was 1964, and I was living at 12,500-foot elevation in La Paz, Bolivia, fulfilling my military obligation as a US Public Health Service (USPHS) physician assigned to the newly created Peace Corps. During that era, all US citizens graduating from medical school were obliged to a minimum of two years of military service or its equivalent, and my hope was to join the navy's program to train flight surgeons because it offered the opportunity to learn to fly. During my internship at Charity Hospital in New Orleans, I reported to Keesler Air Force Base in nearby Biloxi, Mississippi, for a preentry physical examination, my head full of visions of starting flight training the next year at the Pensacola Navy Air Base about one hundred miles to the east. But becoming a flight surgeon was not to be. I failed the physical examination because of a high-frequency hearing loss, of which I was unaware. It was likely caused by shooting guns as a teenager.

And so, encouraged by medical-scientist acquaintances, I then applied to enter the Epidemic Intelligence Service (EIS),

10. Ibid., 98.

11. Ibid., 98.

based in Atlanta at the Center for Disease Control (CDC), as a USPHS officer. Accepted into the program and near the end of my internship, I attended the annual EIS conference at which assignments are made for the incoming EIS class. My assignment to the Kentucky State Health Department was fine with me. During the final day of the conference, however, I was fascinated by a presentation given by an EIS officer who had just returned from a year in West Africa as part of CDC's support for the massive and successful effort to eradicate smallpox from the world. During that year he had worked closely with the Peace Corps. I made it a point to speak with him after his stimulating presentation. He urged me to call John Cashman, the medical director in Washington because, he said, USPHS medical officers were actively being recruited to serve as Peace Corps physicians. With the spontaneity of youth, I followed his suggestion. I was asked by Dr. Cashman to report to his office the next day.

I then grappled with a major decision in my young professional life: EIS or apply for Peace Corps. The latter opportunity being too appealing to pass up, I then faced the fearful task of meeting with CDC's famous and powerful Dr. Alex Langmuir, founder and director of the EIS. I may be the only presumptive EIS officer ever to ask his permission to opt out of this distinguished group of medical detectives. He seemed shocked but agreed for me to explore the Peace Corps option. Having almost certainly burned my bridges with Dr. Langmuir and the CDC, I then faced exhaustive interviews in Washington. My final hurdle arrived the second day, when I was scheduled to report to the office of Sargent Shriver, Peace Corps director and brother-in-law of President Kennedy. But shortly before my meeting with him, I learned that Shriver had been called away and that I should return to New Orleans and wait to be notified when to return to Washington. Those weeks of waiting were excruciating because (1) an option to return to the EIS seemed very dubious and (2) a persistent rumor circulated that Shriver's interviews were extremely rigorous, with only about 50 percent getting past this ultimate hurdle. Somehow, I did make it through, and knowing I was a Texan, Shriver made it a point to introduce me to his then deputy director, Bill Moyers, later Lyndon Johnson's closest assistant and now one of our most respected journalists. I left Washington feeling relieved and supercharged.

Born and reared in San Antonio, Texas, and having studied Spanish in high school and college, my knowledge of Spanish led to an assignment in Bolivia. A physician was urgently needed there, and additional months of language training thus could be avoided. In September 1963, following a few weeks of orientation and training in tropical medicine, I reported for duty in La Paz. With another physician, I was responsible for the health and well-being of over two hundred volunteers in one of the most impoverished nations in the hemisphere. My newly acquired responsibilities were vastly different from my days as an intern at Charity Hospital in New Orleans. Those for whom I was responsible were mostly scattered in rural communities in an area the size of California and Texas combined, and the terrain ranged from towering Andean peaks to vast savannahs and tropical rain forests of the Amazon basin.

Months flew by, packed with the rich experiences and challenges of living and working in this Andean nation. Injuries and health problems were common. One volunteer acquired a potentially deadly tropical parasitic disease (Chagas disease) from a "kissing bug" common in her region. Others were seriously injured in road mishaps and later were evacuated to the United States for treatment. I traveled by horseback to a remote village, far above the tree line, to combat an outbreak of louse-borne typhus. With Bolivia on the brink of civil war, with the Bolivian army surrounding the country's largest tin mine, and with foreign hostages (including several Americans and a Peace Corps volunteer) held by militant miners, I volunteered to travel to the mines with a negotiating team from La Paz. My role was to evaluate the medical status of the hostages and to be nearby to "pick up the pieces" should fighting erupt. Happily, this crisis ended peacefully.

On the streets of La Paz one day, I was astounded to see my friends from Panama, doctors Karl Johnson and Ronald MacKenzie. Both were physicians/scientists employed by the US National Institutes of Health, valued colleagues, and professional role models. They had recently arrived in Bolivia to investigate a mysterious disease, later known as Bolivian hemorrhagic fever, that was slowly killing off the population in and near the town of San Joaquín in the remote Beni region. Their team's subsequent research, conducted with other American and Bolivian investigators, including Peace Corps volunteer nurse Rose Navarro, led to the discovery of a new virus (Machupo). The team also determined how to interrupt virus transmission to humans, thereby saving countless lives. This was one of the epic disease-control triumphs of the century, and while I spent only one week in San Joaquín during the epidemic, the experience provided a profound introduction to the medical/scientific field of epidemiology (crudely defined as the study of epidemics). Karl Johnson went on to become one of America's most distinguished medical scientists through his vital contributions to the knowledge and control of other hemorrhagic fevers: Lassa fever, Korean hemorrhagic fever, and Ebola (which he named).

On November 22, 1963, I was seated in a crude barber's chair in a tiny remote hamlet when a shortwave radio chirped out, in Spanish, something about President Kennedy being shot. It was not until I returned to La Paz two days later that I learned the reality of that chirp. In front of almost every home in La Paz, a flag hung at half-mast. The outpouring of grief and expressions of condolences to Americans were overwhelming. In a distant Andean nation, Kennedy and his creation, the Peace Corps, offered far more hope to people than I had imagined.

Months later, I was unexpectedly assigned to Bogotá, Colombia, for temporary duty when my superiors in Washington wanted me to cover the duties of a Peace Corps physician who was away for two months. Since 1961, Christopher Sheldon had been Columbia's first Peace Corps director, recruited on the recommendation of Mary Bunting, president of Radcliffe College and mother of a student survivor of the sinking of the *Albatross.*

Chris commanded the respect of his staff and the volunteers, but few seemed comfortable approaching him. He was highly qualified for the position by virtue of his impressive academic credentials, his experience as a leader of young people, his fluency in Spanish, and his recent history of having survived the loss of the *Albatross.* Deeply immersed in the challenge of initiating a Peace Corps program that grew into the largest in Latin America, he channeled his energies into the task. With some seven hundred volunteers, it was one of the early Peace Corps's proudest achievements.

Perhaps less intimidated by Chris than were other members of his staff, I enjoyed a comfortable relationship with him, socially as well as professionally. One weekend in Bogotá, we attended a chamber-music concert performed by I Musici, an internationally acclaimed Italian chamber orchestra on tour. The concert was all the more memorable by virtue of its venue, Bogotá's historic Teatro Colón, and the musical program (works of the great Baroque composer Antonio Vivaldi).

While my modest interaction with this inscrutable, solitary man over a two-month span seemed to engender mutual respect and perhaps a hint of friendship, I was still surprised when, on one of my last days before returning to Bolivia, he shared with me his longing to replace the *Albatross* with a similar large sailing vessel and to circle the globe, much like his famed mentors Irving and Electra Johnson had done repeatedly with the schooner *Yankee* and the brigantine *Yankee.* Chris explained that while serving as the brigantine's first mate on a circumnavigation, he had met Alice, his future wife, a physician and ship's surgeon. But I was dumfounded when, explaining that he would need a physician if he succeeded in finding a new ship, he asked if I would be willing to join him in this capacity. While my reply (without hesitation) was affirmative, I was aware that a year or so remained in my service obligation and that innumerable obstacles stood in the way of his resolve to return to the sea. I also faced professional and personal decisions, because I had been accepted into Tulane's surgical residency program and was expected to begin in July 1965. Also, there was the matter of a girlfriend perhaps anticipating matrimony when I completed my Peace Corps service.

After I returned to Bolivia, Chris and I maintained contact via occasional mail and rare phone calls. Thus, I was aware of his continuing search for an appropriate sailing vessel, including inspecting one for sale in Corpus Christi, Texas.

But the leap forward that permitted Chris to purchase the *Verona* occurred in Miami and indirectly involved a notorious and flamboyant character, Captain Mike Burke, who died in May 2013 at age 89. His story is illustrative of the enormous challenge of providing open-sea sailing experiences (short or long) in a safe, ethical, and meaningful manner. Starting in the late 1940s, Burke had created Windjammer Barefoot Cruises in the Caribbean. Marketed to young, affluent folks of both sexes seeking the romantic thrill of comfortable sailing and heavy partying in the Caribbean for a week or two, the popularity of these cruises grew until Burke had a fleet of six sailing vessels that avoided US Coast Guard inspections by sailing only out of foreign Caribbean ports. Crew members invariably lacked requisite training and seamanship expertise and received minimal salaries, if any at all. While Burke was infamous and distrusted by reputable mariners on the Miami waterfront, his business thrived for years until he and Windjammer Barefoot Cruises repeatedly experienced disaster.

The first disaster involved the famous and beloved brigantine *Yankee*, formerly owned by Irving and Electra Johnson. Under Irving Johnson's command, the *Yankee* had completed four highly successful global circumnavigations with amateur crews, and Chris had served as first mate on the last of these voyages (1956–58). Featured in a *National Geographic* TV special in 1966, this vessel and the voyages with the Johnsons were known to millions of Americans. Shortly after the Johnsons sold the *Yankee* to a man who used it for summertime sailing off the coast of New England, it was purchased by Burke. He then hired a skipper and crew, who, on February 10, 1964, sailed from Nassau on a scheduled fourteen-month global circumnavigation. One of its passengers disappeared mysteriously in the Galápagos Islands and was never found. Leaving Ecuador abruptly without paying the levied police fine, the *Yankee* sailed on to the Marquesas and Tahiti, but after months of disarray and poor maintenance and shortage of funds, the *Yankee*, on July 24, 1964, ended up aground and abandoned on a reef off Rarotonga, Cook Islands.

Worse yet, in October 1998, another of Burke's vessels, the schooner *Fantome,* was lost to Hurricane Mitch near Roatán in the Caribbean. All thirty-one aboard perished. Curious readers may be inclined to seek additional information about the life and times of Michael Burke.

After months of searching, Chris heard about a barquentine named the *Grand Slam* that had just been put up for sale in Miami. Owned by a wealthy German attorney whose fortune stemmed from post–World War II reparations, the *Grand Slam* had set out from Europe with a young, affluent crew, many of whom had titles connecting them to noble families on the continent. Their plan to circle the globe leisurely for about four years was interrupted when, upon reaching the Caribbean, the owner suffered a heart attack. After he recuperated and rested for a month or two, the *Grand Slam* continued its voyage, only to be forced again to stop in

Miami when the owner suffered a second heart attack. He was hospitalized in Miami, and the vessel, manned by a skeleton crew, was offered for sale. The 350-ton vessel, built in the United Kingdom in 1912, was 138 feet in length, had a teak hull and decks and three masts (with square sails on the forward mast). It was one of those proud and magnificent seagoing sailing vessels that seemed to emerge mysteriously from an earlier era. And of course, Mike Burke was determined to purchase it.

To his immense credit, the German owner refused to sell the *Grand Slam* to Burke because of his notoriety, in part related to the then-recent grounding of Burke's *Yankee.* Chris was thus able to purchase this exceptional vessel for about half the original asking price.

After hearing from Chris that he had purchased the *Grand Slam* (and would rename it the *Verona*), my elation was balanced by the pressing need to advise my parents of my plan to spend a year at sea instead of entering a surgical residency in New Orleans, request a year deferral from the Department of Surgery at Tulane Medical School, share the plans with my girlfriend, acquire new photographic equipment, and complete myriad other tasks. Fortunately, much of my USPHS salary remained in a savings account, and my significant medical-school debts would not become due for many years.

In retrospect, my parents were likely aware and accepting of my rather peripatetic nature but would have preferred postdoctoral medical training in the United States to sailing off into the unknown for a year. Perhaps they remembered that when I was a child, *National Geographic Magazine* fascinated me, its photos transporting me to distant lands and cultures. And as a youngster, the opportunity to travel was a magnet, whether driving to Ohio with a cousin just discharged from the navy after World War II, attending a Boy Scout jamboree in Valley Forge, Pennsylvania, in 1950, camping in Big Bend National Park as an Explorer Scout, or accompanying my family to Mexico City. My mother and father did not have the opportunity to travel abroad until much later in their lives, but a modicum of wanderlust must have been part of their makeup. After all, within years of their marriage in Cincinnati in the 1920s, they had moved to the wilds of Texas. My father, a newly graduated optometrist, set up practice in Hallettsville, Texas (midway between Houston and San Antonio). My brother, Jerry, was born there just fifteen months before my birth, in the Nix Hospital in San Antonio. Family lore has it that the Ohio families thought that the newlyweds were behaving irrationally by heading off to the wilderness of Texas, an uncivilized destination.

Although the German owner of the *Grand Slam* had spared no cost in outfitting the vessel, transforming it into the *Verona* required considerable reoutfitting. And the crew needed to be recruited before leaving from the Bahamas in June 1965. Chris resigned his Peace Corps position and reactivated the Ocean Academy Ltd. He placed advertisements in yachting and sailing magazines and in *National Geographic Magazine.* Also created were postcard-sized cards with a fine color photograph of the *Verona* at sea.

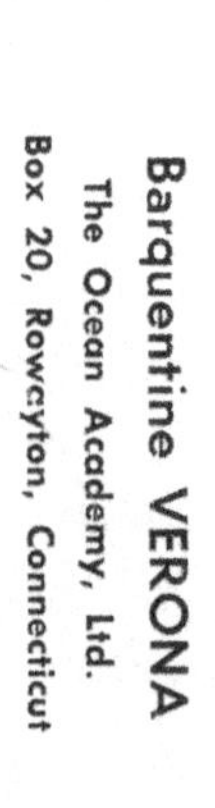

Figure 4. Card showing the *Verona*'s upcoming voyages.
Created by The Ocean Academy, Ltd.

The back of the card described the Barquentine *Verona*, scheduled voyages, and crew eligibility for applicants.

> BARQUENTINE VERONA 138' on deck, 24' beam, 13' draft, displacement 350 tons
>
> The Barquentine VERONA is one of the finest sailing vessels of her size in commission in the world today. Her hull is three inch teak planking bronze fastened to iron frames. Virtually all her equipment-twin diesel engines, twin generators, wiring, tanks, electronics, accommodations, etc., are new since 1962. She carries over 8,500 square feet of hand sewn Dacron sail.
>
> The VERONA is under the command of Christopher B. Sheldon, Ph.D., a man with extensive experience in all kinds of sailing vessels. He has navigated more than 100,000 miles under sail in the major oceans and seas of the world. The vessel normally carries an additional senior crew of First Mate, Engineer, Doctor, and Cook. The rest of the crew is made up of amateurs who carry out the full functions of seamen.
>
> Each year the Barquentine VERONA makes eight to twelve month voyages to distant parts of the world. Scheduled are:
>
> > June 1965–June 1966: World voyage through the South Seas, Indian Ocean and Mediterranean.
> >
> > June 1966–May 1967: African voyage from Lisbon, Portugal, through the Mediterranean, Red Sea, around Africa, and across the Atlantic to the Caribbean and United States.
> >
> > July 1967–July 1968: Nations of the Atlantic—Scandinavia, Holland, England, Portugal, Brazil, Argentina, South Africa, West Africa and the Caribbean.

> The voyages are open to young men, women, and married couples approximately 16–35 in age who share the expenses and work as full crew members.

Many of the crew members first learned about the *Verona* from an advertisement in the *National Geographic Magazine.* Its 1965–66 voyage departed from Nassau, Bahamas, in early June, lacking only the ship's surgeon until I joined them on June 26 in the Canal Zone. Dr. Karl Johnson, then director of the Middle America Research Unit in the Canal Zone, transported me to dockside, wishing me safe travels as I hauled my possessions onboard and into my assigned cabin in the forecastle. My last meal on shore included apple pie à la mode and a cold beer with Karl and his wife, fellow scientist Dr. Patricia Webb. The pilot boarded at 16:00, and we weighed anchor an hour later. Our destination was the Galápagos Islands, some two weeks distant, just below the equator in the South Pacific. The *Verona*'s size and beauty attracted much attention as we were literally circled and escorted by Panamanian speedboats filled with young people waving and asking our destination and wishing us safe travels. The thought occurred to me that, save for my sail mates, the enthusiastic faces of the speed boaters were the last I would see for weeks.

My new job as seaman and ship's surgeon under the command of Captain Sheldon had begun as Balboa, Canal Zone, slowly drifted from sight. Assigned to the 04:00–08:00 and 16:00–20:00 watch, I was just beginning my crewing duty when at 16:00, Chris barked orders to set all the fore and aft sails and to steer a course of 190 degrees. I had never been on a sailing vessel of this size and complexity, and although I managed to memorize much as a medical student, I was seriously challenged to learn the names of all of the *Verona*'s lines and pulleys and other nautical terms needed to quickly follow the orders of the watch captain. There was also the acute need to become comfortable with working aloft.

Surprisingly, the initial fear of climbing twenty to thirty feet above the deck, masts swinging side to side and loss of grip promising severe injury, if not death, subsided quickly. Over time I found that working aloft, concentrating on the task at hand, was among the most pleasant and memorable moments at sea. And happily, I was able to fend off seasickness, not an uncommon source of misery suffered by some crewmates.

Figure 5. View from aloft. Photo by author.

William "Bill" Bunting was our first mate. As a survivor of the *Albatross*'s sinking, Bill enjoyed unique status among the crew. His special relationship with Chris was perhaps a source of envy for some. Bill had been immersed in things nautical most of his life and was a consummate mariner who, despite his youth, was held in high regard by the crew. He rarely left the *Verona* during the entire year; indeed, in port he could almost always be found on deck or aloft doing intricate repairs or maintenance. Not surprisingly, as an academic, he is now the respected author of a number of books on nautical topics.

My first medical call came on the second day out of Panama, when Edgar Faust suffered a nail wound on his left foot that required tetanus toxoid. An air of excitement stirred on deck later that afternoon when a five- or six-foot shark was caught on a line, and for some reason, crew member Tom McMasters decided to shoot it with his rifle.

Curiously, days later, the weather cooled as we approached the equator. Weather conditions in this region are influenced by the cold Humboldt Current, which issues forth from the Antarctic, paralleling much of the west coast of South America. Our proximity to the equator jolted me into the realization that dramatic new experiences lay before me.

Two

Birds and Coffee Beans in Honduras

The United Airlines Boeing 737 carrying us from Houston to Honduras was nearing touchdown when I told my wife that I had just read in the *New York Times* that our destination, San Pedro Sula, had the world's highest homicide rate. Adventuresome and curious but fortunately less impulsive than I, Nancy responded with a smile of sorts and a comment to the effect, "Now you tell me." The year was 2012, and we were off to meet my former *Verona* crewmate Lloyd Davidson. Forty-seven years had passed since seeing him. A bus would take Nancy and me to Copán Ruinas, where Lloyd lived, a bus ride of some six hours to the west, toward Guatemala.

The trigger to my quest to find Lloyd was a photograph. During my year at sea, I had taken thousands of photographs, mostly slides. They remained well preserved and unseen in boxes, transported for decades from one home to another, until I retired, and the digital age had matured. Finally, I had time to methodically review the slides, select over three hundred "keepers," and convert them into digital images stored in my computer. The images worked their magic, transporting me instantly to places where I relived long-past moments of discovery. Many images reconnected me with forgotten deposits into my experience bank, and they started me seriously thinking about capturing on paper highlights of that magical year.

Starting with photos during the *Verona*'s passage from Panama to the Galápagos Islands, memories of my first days at sea were rekindled. A photo of young Lloyd Davidson at the wheel caught my attention because of all the young men on the *Verona*, Lloyd had impressed me most. It was late June 1965 when I first met him. He was on duty at the helm, fiercely focused on maintaining the designated compass setting. With trailing winds, it was a challenging task demanding concentration, and for which failure could be rewarded with a sharp dressing down by our skipper.

Figure 6. Lloyd Davidson at the wheel. Photo by author.

Upon meeting, we exchanged but a few words. Lloyd was nineteen years old, a native of Knoxville, Tennessee, and had just completed his sophomore year at Davidson College (no relationship) in North Carolina. For several reasons, Lloyd and his Knoxville childhood buddy, Edgar Faust, stood apart from the other young men in the crew. First, in exchange for setting the mess table, transporting food from the dumbwaiter to the table, and clearing the table and washing dishes, they each received a 25 percent reduction in the cost of their year at sea. Second, there was a mannered purposefulness about the two of them, a tribute to their Southern upbringing, which distinguished them as young gentlemen, if such a thing is possible on a long ocean voyage. Third, and most important, they were acknowledged masters of their chosen sports: fishing, spear fishing, scuba diving, swimming, or anything else to do with denizens of the deep. Their knowledge, skill, and physical conditioning were hard-earned. From an early age, they had spent their summers at a skin-diving camp on the Caribbean Islands of Trinidad and Tobago. During their midteens, they had transitioned from campers to staff, providing ample opportunity to expand and strengthen their skills and their passion for the sea.

There were twenty-five of us on the *Verona*. A neutral observer might be tempted to divide us into two broad categories, sailors and travelers, with more or less equal numbers of each. The former were drawn to the *Verona* by the desire for a global seafaring experience. The latter found the *Verona* an inviting and exotic vehicle to take them to distant, relatively inaccessible lands. An artificial dichotomy, perhaps, not intended to suggest that Lloyd and Edgar lacked interest in our ports of call, but these guys' primary passion was clearly linked to the sea. Whenever we approached landfall near coral reefs, the two of them would not be found pouring over tourist guides but rather checking their masks, filling tanks with compressed air, inspecting regulators, and sharpening spearguns. For days at a time, they were literally underwater, not visible except for brief surfacings. They consistently earned the goodwill of their fellow crew members by providing the freshest of choice fish for our dining pleasure.

Looking at Lloyd's lean, confident image at the *Verona*'s wheel, scuba-diving watch on his left wrist, led me to wonder what had become of him since our crew dispersed in Lisbon in June 1966. So I did what had become a reflex by then. I googled him. This simple act led to a prolonged series of communications that eventually led me and Nancy to board the flight to San Pedro Sula to reconnect with him face-to-face. This is how I remember it happening. Online browsing led me to the description of a bird sanctuary that appeared to be Lloyd's. It was located in the town of Copán Ruinas, Honduras, and it had an e-mail address. Believing my search would be quick and easy, I immediately sent a message expressing my hope to exchange e-mails. I anticipated a quick reply, but weeks passed. Nothing. Disappointed, I renewed my online search and found a travel agent's page that described in glowing terms her recent visit to the bird sanctuary.

This helpful travel agent (whom I later learned lived in Austin, Texas, an hour from my home), after listening to my quest, offered to put me in touch with Flavia Cueva, owner of Hacienda San Lucas, near Copán Ruinas. Flavia was described as a close friend of Lloyd's. And what a charming lady Flavia turned out to be! Confirming their long-standing friendship, she explained via e-mail that Lloyd was not prone to check his e-mails regularly and that Wi-Fi service was rather spotty, but she assured me that she would deliver my message to him. Sure enough, a few days later I received an informative e-mail from Lloyd explaining that he had been living in Honduras for decades. Encouraged by his good response and curious to learn more, I offered to travel to Copán Ruinas. Flavia (by this time, we had bonded via the Internet) suggested October as a good time to visit and was pleased to book us into one of her ten guest rooms at Hacienda San Lucas. She also offered useful tips about transportation options (a reliable, comfortable five-to-six-hour bus trip) from the San Pedro Sula airport, and she sent a driver to take us the final few miles from the bus station to Hacienda San Lucas.

Flavia is the prototypical perfect hostess—and a lady with a captivating story. Born into a prominent family of the capital city, Tegucigalpa, she barely tolerated the frequent and boring long family drives to visit her grandfather's working hacienda near Copán Ruinas. She was a city girl for whom the remote mountains of western Honduras offered little appeal. Headstrong and highly intelligent, Flavia managed to get herself expelled from Tegucigalpa's elite girl schools, attended college in Kentucky, married, and lived there for forty years...working as a teacher and as a caterer until, widowed, she returned to her native land. By this time, she had inherited her grandfather's inactive hacienda, which lay in ruins and had little appeal to other family members. With the help of a much younger lover from a social class far below hers, she not only shocked her family and the community, but she also managed to restore the hacienda to receive paying guests. And her culinary skills soon created the most desirable dining destination in the region. Encouraged by her friend Lloyd, Flavia expanded the hacienda from four rooms to ten, and Hacienda San Lucas became known as a prime refuge offering creature comforts in a jaw-dropping rustic tropical setting overlooking Copán Ruinas and the monumental remains of the Mayan complex of Copán.

After our arrival, and with Flavia as our bartender and gracious hostess, Lloyd suddenly appeared. He was agile and fit and seemed decades younger than his chronological age. Magically, the years dissolved as we shared glimpses of our lives during the previous forty-seven years. Moving from the rustic bar to the spacious veranda, our eager conversation reduced an otherwise delectable dinner to secondary importance.

Lloyd and his childhood buddy Edgar, having convinced Davidson College's top officials to grant them a year of leave to crew on the *Verona*, returned as promised to complete the final two years of their undergraduate education. The war in Vietnam was escalating when they graduated in 1969. Lloyd joined the navy and Edgar the coast guard, each serving for five years. Most of their service time was spent at sea, with Lloyd's final years served as a hard-hat navy diver on an icebreaker assigned to the Arctic Circle. Afterward, they reconnected with the friend who still operated the skin-diving camp in Trinidad and Tobago, and Lloyd served on the staff until he and Edgar (by now, both had wives) decided to realize their persistent dream of more or less retracing the *Verona*'s circumnavigation of the globe in their own vessel. They accomplished this feat over three years in their forty-eight-foot ketch—more remarkable considering that these master mariners eschewed all electronic and navigational tools except for a handheld sextant! I have been privy to mere snippets of that remarkable adventure. Facing economic challenges at times, their mastery of ships and ship repair opened windows of opportunity to supplement their incomes with cash or bartered goods. So the three years were dotted with a variety of months-long stays in waterfront settings where they practiced their trade. After three years, they and their sailboat returned safely to the Caribbean, but by that time, Lloyd's marriage had fallen apart.

Cashless, Lloyd was willing to return to his skin-diving camp position, now as director, but for complex reasons, political and financial, the camp had been moved to the island of Roatán, off the coast of Honduras. Lloyd went to Roátan, where he and wife number two lived as expats. At this time, birds entered Lloyd's life. It seems a common island phenomenon: expats arrive, fall in love with the magnificent local parrot species (especially toucans and macaws) and other avian species, and acquire one or more feathered friend. But typically, the expats tire of the pet bird, or of the spouse, or of island life, and depart, necessitating that a home be found for their bird or birds. Lloyd's wife was the go-to bird lady of Roátan, adopting abandoned parrots until their number exceeded fifty. At some point, however, it appears she decided to escape the birds, or Lloyd, or island life, or all of the above, leaving him with an abundance of coddled winged friends. By now Lloyd had become very fond of the birds, and had learned a great deal about their biology, ecology, and care.

After moving to Roátan, Lloyd had not remained cashless for long. While operating the skin-diving camp, he recognized the absence of commercial-scale fishing in Honduras, a Central American country with over eight million inhabitants and an extensive Caribbean coastline and a small Pacific one. So he set out to create this industry, with noted success. Over the years, he became a prominent, respected businessman in Honduras. He also had developed a close relationship with a charming American businesswoman, Michele Braun, a Roátan resident, artist, and owner of a popular art gallery and gift shop.

During our first dinner, and over the exquisite cuisine that emerged from Flavia's kitchen, Nancy and I tried hard to assimilate details of Lloyd's engrossing account of decades of his life. One of us mentioned that we had friends in San Antonio, Jim and Gwyn Creagan, who had lived in Tegucigalpa. Jim, a retired career Foreign Service officer, had served as US ambassador to Honduras. Lloyd smiled and said, "Yes, I remember them very well. Gwyn used to call me when she needed choice fish for diplomatic gatherings." And when we next saw the Creagans and mentioned our visit with Lloyd, Gwyn acknowledged their friendship with a broad smile. Yes, Lloyd's fish had contributed to many a successful dinner party in the capital.

After ten years, Lloyd abandoned his commercial fishing enterprise because he realized that his fishing-boat captains could (and did) profit more by transporting drugs than fish. As he gradually withdrew from the Roátan world, he started to plant roots in the lovely western mountains of Honduras. This process was greatly assisted by the fact that Michele owned a house in the town of Copán Ruinas and was very familiar with the lay of the land in this splendid area. Adjacent to the town are the southernmost of the major Mayan ruins, Copán.

Prolonged, complex negotiations with a local landowner finally yielded Lloyd's prized acquisition of several acres of prime land along a pristine river near the center of town. After being meticulously and lovingly groomed to become a bird sanctuary, this land evolved into Macaw Mountain, populated initially by the fifty or so parrots flown from Roátan to Guatemala and then transported overland to their lush destination. Macaw Mountain grew into a popular tourist destination, with peak tourism occurring around the time of the highly publicized Mayan prophecy of the end of time. A guy who thrives on challenges and does not fear reinventing himself, Lloyd acquired impressive knowledge of the ecology and protection of local parrots and then created a strong, proactive educational program for schoolchildren and the public at Macaw Mountain. He has also undertaken, in partnership with national authorities and local experts, efforts to reintroduce and restore the free-flying population of Macaws (sacred to the ancient Mayans) to the grounds of the vast Mayan temple complex and elsewhere.

As our leisurely, informative tour of Macaw Mountain ended one afternoon, Lloyd asked if we would like to visit his coffee plantation. Of course we accepted this offer, having no idea what to expect. Remarkably, we learned that in a period of a few years, Lloyd had managed to acquire vast knowledge of the array of challenges faced by coffee growers and producers. Learning more about coffee (and Lloyd) throughout the delightful day at his plantation, Cafe Miramundo, we were also treated to magnificent mountaintop panoramas when they were not obscured by passing clouds. In Lloyd's beautiful and cozy home, we enjoyed a bountiful, gorgeous lunch prepared by Michele, who had just spent several grueling hours on her mountain bike, training for an upcoming biking trip over the Andes from Argentina to Chile. As Michele prepared lunch, Lloyd described the multitude of obstacles he faced after purchasing the plantation. He then walked us through the equipment used to separate the finest, perfect beans from the others, explaining in detail what took place in this area when the coffee harvest was underway. Cafe Miramundo is not only one of the highest quality coffees sought globally but is produced according to the stringent standards of ecological, social, and financial practices set forth by international agreements.

Figure 7. Lloyd and Michele at Cafe Miramundo house. Photo by author.

Our final treat of that special day at Lloyd's coffee plantation was a downhill stroll on a path that took us to a picture-perfect waterfall cascading into a natural pool used for swimming and relaxing. But for Lloyd, there seemed to be little of the latter. He is no idle, privileged, absentee plantation owner. Much of what he showed us with quiet pride was created with his sweat equity, working side by side with his loyal, well-treated employees. One story exemplifies his relationship with an employee, his Cafe Miramundo foreman, a local man born in a nearby village in these very mountains. In addition to building his own very comfortable, modern, and substantial home, Lloyd built another somewhat smaller home for his foreman. Both homes had modern kitchens and appliances. After the foreman and his family occupied their new home, Lloyd observed that they spent very little time in it. They constructed a modest open shelter (lean-to), with an open fireplace where they did all the cooking. When asked, the foreman confessed that the family was not very comfortable living in their new home as it did not fit their traditional ways. Cooking on a modern stove top was simply not to their pleasure. So Lloyd, bilingual and fully attuned to the local culture, agreed to build another home, a "simple" one with an open outdoor fireplace, which perfectly fit the needs and desires of the foreman and his family. And, always a good businessman, Lloyd remodeled and expanded the foreman's former home into a comfortable "lodge" that can sleep up to eight guests. It is available to renters who seek exceptional mountain beauty and solitude less than an hour's drive (four-wheel drive essential) from Copán Ruinas.

I keep in touch periodically with Lloyd and Michele and with Flavia by e-mail, itching to return to their seductive tropical world and to add to my appreciation of this Knoxville native. Possessing abundant charm and intelligence, and having twice circled the globe, Lloyd navigated his way to a Central American republic that benefited significantly from his presence (and vice versa).

Nancy, good traveler, keen observer and excellent judge of people, shares my enthusiasm about having reconnected with my old crewmate. And an unanticipated bonus resulted from our trip: Nancy heard Lloyd spontaneously relate accounts of nearly half-century-old shared implausible adventures, using almost identical words and phrases she had heard from me during decades of our married life. A credibility boost for me at the very least.

Three

Surgery by Flashlight

From his mountain perch, Luis Carvajal scanned the horizon and was elated to see a white sail appear. It was his last hope. Mariana, his eleven-year-old daughter, was desperately ill, her condition worsening by the hour. She was almost delirious with fever, unable to eat for days, and too weak to speak beyond a whisper. His prayer was that the sail was attached to a vessel large enough to carry a doctor. Encouraged as it approached Santa Cruz Island, Luis gently loaded Mariana onto his sturdy mule and began the agonizing descent to Academy Bay.

During moments of leisure at sea, far from the sight of land, I was sometimes seized by visions of denied pleasures. My recurrent fantasy was to be comfortably seated, leisurely sipping a beer cold enough to chill my throat and cool my hand. But because our floating home was an alcohol-free "dry ship" (or so I thought during the early part of our voyage), and because our first destination after Panama was about two weeks distant, I had no choice but to wait. Imagination was my only access to that forbidden treat. On July 7, 1965, we reached the Galápagos (Ecuadorian) island of Santa Cruz Island and dropped anchor in Academy Bay.

Moments away from that dreamed-of sip, my anticipation was interrupted by a shoulder tap. Luis Carvajal's immediate question for me was "Is there a doctor on your boat?" I wetted my lips, still standing at the primitive shack, which offered...yes...cold beer, and answered, "Yes, I'm the doctor." Having lived in South America during the previous two years, I was fluent in Spanish. The anxious father told me about Mariana's illness and asked if I could see her. Immediately, we took her to a tiny municipal facility and laid her frail body on a table. Acquiring more information about her medical history and examining Mariana, I found a large, warm, and tender mass on her lower-right abdomen. I learned that her illness had begun five days earlier when she felt ill and lost her appetite. Mariana's condition worsened over the following two days, and the pain had become more severe. Then a mass had appeared as her condition further deteriorated. Suspecting a ruptured appendix, I could not rule out other causes. She was febrile and barely responsive. Her condition was precarious. Mariana's temperature was over 103 degrees, and she was severely dehydrated.

I hurriedly returned to the *Verona* to collect the rather meager medical equipment and supplies available to me: antibiotics, syringe and needle, local anesthesia, a scalpel and blade, bandages and tape, forceps, suture needles and suture material. Anticipating that a drain might be needed, I also grabbed a piece of rubber enema tube that I chanced upon. Because we were near the equator, where night falls suddenly, I also took along two flashlights. It was dark before surgery could be initiated; flashlights complemented the lone dim bulb in the room. In addition to the calming presence of Mariana's astute and resolute father, local residents Jim Perez, Bud Devin and Gus Angemeyer assisted in immobilizing Mariana as I prepared her for surgery. Using minimal local anesthesia and picking up the scalpel, my thoughts raced through possible outcomes and considered the potentially deadly prospect that I might be opening a loop of trapped bowel. But to my overwhelming relief, copious quantities of yellow pus flowed from the incision, confirming the presence of a huge abscess. Fortunately, her inflamed appendix had ruptured against the abdominal wall instead of into the peritoneal cavity. Using surgical principles I had learned in my medical training, I knew that a rubber drain would help prevent a rupture and the peritonitis that might prove fatal. I fashioned a drain out of a strip cut from the enema tube and secured it firmly with sutures.

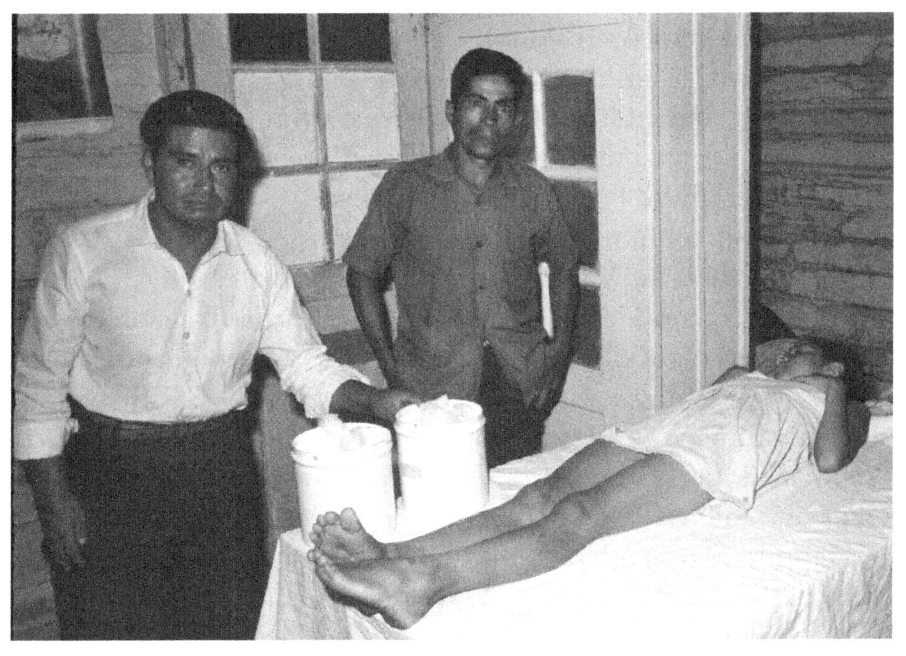

Figure 8. Mariana immediately before surgery;
father Luis Carvajal at left. Photo by author.

During the procedure, a stoic Mariana hardly moved, nor did she cry out. Luis held his daughter's hand and spoke softly near her ear. After giving her antibiotics and urging her to sip liquids, there was little more that I could do but wait. Mariana remained on the same table overnight, with Luis by her side. After a restless and painful night, in the morning, Marianna looked better, but her temperature was still 102 degrees. A very positive sign: her normal bowel sounds and a soft belly indicated that the surgery may have prevented rupture of the abscess into the abdominal cavity.

The *Verona* had been scheduled to sail that day to Genovesa Island, but Chris readily agreed to remain at Academy Bay so that I could monitor Mariana's progress. By the following morning, July 9, Mariana had improved considerably, and her temperature was coming down. That afternoon, her fever was almost gone, and she was alert and eating a bit, and her condition continued to be favorable. I felt that at this point she was likely out of danger and removed the drain. Promising Luis that we would return within a week to check on her, we sailed at midnight.

We returned to Academy Bay the morning of July 15. I was eager to check on Mariana and to visit with friends made on our previous visit, especially with Forrest Nelson, an American seafarer who had fallen in love with this remote speck in the Pacific Ocean, sold his sailboat, and started building a small hotel. Forrest was also an avid ham radio operator, and he kindly offered to patch a telephone call to the United States for me.

Happily, Mariana had continued to improve. Now I found her in excellent condition, walking, pain-free, and eating normally. Her surgical wound was clean and healing nicely, allowing me to remove the remaining sutures. Her thankful father asked when we were departing because he had a gift for me. As we boarded the *Verona* several hours later, I looked up to see Luis Carvajal arriving with a cart laden with watermelons, papayas, oranges, lemons, bananas, avocados, and even home-made cheese. He gave me a big hug, and I vividly remember his smile and wave as our anchor was raised, and we glided toward our next destination, Isabela Island.

Figure 9. Loading Luis Carvajal's gifts onto the *Verona*. Photo by author.

About five months later, a letter dated November 25, 1965, arrived at my home address in San Antonio, waiting for my return. It was written in longhand in English, certainly by someone other than Luis.

> Dear Doctor,
>
> Are sending our best regards in these few lines, hoping to express our gratitude for all you have done for our Mariana and her parents.
>
> We all have you in dear memory.
>
> We can only wish you a good health, so you may continue helping many suffering people.
>
> The operation gave good results. Mariana has been in school for a long time and is very happy.

> Sending best of wishes for coming New Year from Mariana and her grateful parents, I dare sign friend forever
>
> Luis Carvajal

Decades later, friends Allen and Helen O'Brien travelled to the Galápagos, expecting to be on Santa Cruz Island. I asked them to check into the status of Mariana. They reported that they met her and her children and that they were a healthy lot. I smiled and thought of Luis's gift of a mountain of fresh fruit—and of perhaps a rare instance in which a life might not have continued without my intervention.

Four

Mutiny Mid-Pacific?

The afternoon of September 20, the *Verona* was anchored off remote Palmerston Atoll, one of the northernmost Cook Islands. Although I don't know when the captain's worrisome behavior first became apparent, this is where forces separating the crew and its master acutely collided. On this afternoon, Chris's dangerous abuse of alcohol could not be denied.

Before joining the *Verona*'s crew, I had heard that the loss of the *Albatross* took the life of Chris's wife, Alice, and four students, and that following this tragedy, Chris had manifested a "drinking problem" before his energies became focused on building the Peace Corps in Colombia. But nothing I saw or heard in Colombia suggested that this problem persisted.

When I learned in Panama that the *Verona* was "dry," I accepted the news positively, assuming that experiences from previous voyages on the *Yankee* had taught Chris the wisdom of this policy. And it never occurred to me that anyone aboard was ignoring the rule, least of all he. Chris's large cabin was at the stern of the vessel, far from my forward location, and I have no recollection of entering or even being close to his cabin door.

At sea, we typically did not see much of Chris. Daily at noon, he climbed to the upper deck with his sextant to "shoot the horizon" to estimate our position. Often, on the way to the Galápagos Islands, we lingered near him while he made the necessary calculations to determine latitude, longitude, and distance covered during the previous twenty-four hours. In late July, heading for Pitcairn, Chris offered to teach navigation, and several of us found it stimulating to work on navigation problems he assigned us. His spirit seemed to be buoyed on days such as July 23, when the sextant and his calculations revealed that during the previous day, we had covered an impressive distance of 230 nautical miles under sail.

Figure 10. Captain Sheldon with sextant, mid-Pacific.
Photo by author.

Usually, Chris was reserved, seemingly lost in thought, but after we departed from Pitcairn for the Marquesas Islands of French Polynesia, a long passage with stormy weather and heavy seas for several days, he often appeared irritated, especially in the late afternoons. He would belittle a crew member for little cause or shout at the helmsman for being off course, and exhibit other behaviors that in retrospect were likely fueled by alcohol. Around this time, I noted an occasional hint of spirits such as vodka on his breath. Whether related to the apparent change in Chris's behavior or to the rough seas, I noted in my log book on August 16, "marked increase in discontent and personal conflicts among the crew members." Near midnight on that day, a rather terrifying sequence of events occurred in rapid succession.

Under full sail and moving at a rapid clip, we approached the main harbor at Nuku Hiva with Chris on the top deck, where he normally stood while approaching a destination. With the shore rapidly approaching and no order from our captain to drop sail, it appeared the *Verona* would crash into the shoreline and be destroyed along with whatever we rammed. Senior crew sprang into action by shouting out the necessary orders. Chris, grossly inebriated, became a silent observer to the frantic and thankfully successful effort to prevent catastrophe. The *Verona* was not secured with anchor dropped until after 01:00. Having arrived in the Marquesas without required visas, the local officials boarded the *Verona* about 09:00, and shortly afterward, Chris, still wobbly and with slurred speech, accompanied three gendarmes ashore. By the time he returned in the late afternoon, he had improved, greeting us with the welcome news that the crew was now permitted to go ashore.

The raw and powerful beauty of volcanic, mountainous Nuku Hiva, and its welcoming inhabitants, must have distracted the crew because neither the near disaster nor our captain's incapacitation was discussed. It was on this exquisite island that young Herman

Melville and shipmate Richard Tobias Greene jumped ship in 1842 and were captured by natives. *Typee*, published in 1846 as Melville's first book, relates this extraordinary experience on Nuku Hiva.

Two days after arriving in Nuku Hiva, we moved to the opposite side of the island for dedicated "workdays" anchored in Anaho Bay. Having been denied permission by the authorities to go ashore there, we spent August 20 and 21 cleaning barnacles, scraping paint, painting, and polishing brass. Happily, the following day, the *Verona*'s crew was treated to a traditional Polynesian feast. While this fete was quite enjoyable and introduced us to local cuisine and its preparation, the village's adults seemed more concerned with protecting their daughters than with feeding and entertaining their foreign "guests."

After leaving Anaho Bay, a four-day passage took us to Papeete, Tahiti, where we remained for one week. We then spent almost two additional weeks in the sparsely populated remote Society Islands, where our concerns about Chris had moderated as his demeanor improved. Oddly, in retrospect, it was not until September 16 and while underway to Palmerston Atoll that senior crew began to openly discuss the dire implications of our terrifying midnight arrival at Nuku Hiva. Perhaps I and others had repressed it or had simply feared to deal with it. But it was not until then that I learned and accepted the fact that *Chris had been drinking heavily for quite some time.* The senior crew all agreed that without an intervention our well-being was seriously compromised.

It was the afternoon of September 20 that our captain was asked to join his crew on deck. Obviously intoxicated all that day, and erupting in rage at crew members with minor provocation, something had to be done. Kip Durren, the *Verona*'s engineer and longtime friend of Chris's, took the lead. With the entire crew assembled, Kip confronted Chris with our concerns, explaining that all of us would leave when we reached Hong Kong unless he abandoned alcohol. His response was one of contrition, promising the crew to stop drinking and to throw overboard his entire supply of spirits. While this act was neither observed nor otherwise documented, he did seem to stop drinking, and at least for several months, his demeanor improved dramatically. And not until we were detained in Saudi Arabia (chapter 7) did his rage again erupt. But with one other exception on Santorini, I never again saw Chris inebriated, even on side trips with him.

Was it a mutiny that took place at Palmerston? Or was it a mutiny averted? By any definition, it was certainly an ultimatum issued. I still do not know if Chris's alcoholism was precipitated by his loss of the *Albatross.*

Five

Peripatetic Dreaming I: Galápagos to Formosa

July 2, 1965

Approaching the equator and the Galápagos Islands.

July 10, 1965: Genovesa (Tower Island), Darwin Bay

Log entry: *"Unique for many species of birds, no fear of man. Found roosting in large numbers in low brush and small trees, appearance of a bird-tree. Gulls and frigate birds most common, also blue-footed booby. Bay is half of an intact volcanic crater so one actually sails into the crater."*

July 12, 1965: Santiago (James Island), James Bay

Log entry: *"With Kip and Ginny Durren (unexpectedly) meet American fishermen and invited to board their ship the Santa Anita out of San Diego. Received warmly by Skipper Armand*

> *Tavares. Toured ship, met crew and chief engineer and cook. Served remarkable feast of filet mignon, mushrooms, salad, rice, navy beans, succotash, peas, sour cream, butter, bread, rice pudding, whipped cream, ice cream, and offered good cigar. Fabulous hospitality. Skipper Tavares takes us and some of their crew via launch to Verona, boards Verona, crews visit each other, cooks meet and talk, and our crew invited to Santa Anita for ice cream."*

Foreign ships such as the *Santa Anita* were not permitted to fish in Ecuadorian territorial waters, but laws were ignored because this was a favored fishing area for tuna, and enforcement was virtually impossible. The *Santa Anita*'s crew had been away from San Diego for about two months, and they were delighted to meet and entertain "new faces." Large commercial fishing operations like theirs understand the need to boost morale by providing exceptional meals for the crew. Of course, this was an unexpected treat for the three of us accustomed to the adequate but generally uninspiring food served on the *Verona*.

July 15, 1965: Academy Bay, Santa Cruz Island

We had returned, as promised, to check the status of Mariana Carvajal, and I was delighted to see how she had improved. Her surgical wound clean, I removed the few remaining sutures. Before departing the Galápagos, and waiting to receive Manuel Carvajal's gifts, I paid a final visit to Forrest Nelson and Marlena at their tiny hotel. Forrest and I enjoyed a leisurely few hours chatting, and we promised to stay in touch with one another. This parting was, in retrospect, the first sample of an uncomfortable feeling I was to experience repeatedly during the year—that of being obliged to part with new and valued friends.

July 22, 1965

Log entry: *"Passage to Pitcairn day #6. Longitude 8.30 South, Longitude 108.36 West Distance 24 hours 228 miles. Started navigation classes."*

July 31, 1965

Log entry: *"Arrived noon. Longboats with about 20 men arrived about 2 PM, bearing fruit and other gifts. Sears items they had requested were loaded onto longboats. Two-thirds of our crew went ashore, leaving skeleton crew of 7."*

On and off for a week, we lived with descendants of the mutineers of the *Bounty* fame, several of whom were directly related to mutiny leader Fletcher Christian. After the infamous mutiny in 1789, Christian ordered Captain Bligh and eighteen of his loyalists into a small boat and left them to their fate of almost certain death (which, miraculously, most avoided). Months later, after leaving some of the mutinous crew and seven of the loyalists in Tahiti, Christian, with eight other mutinous crew members and eighteen Polynesian women and men, sailed on the *Bounty* through several Pacific island groups searching for a safe hiding place. Finally, they came upon uninhabited Pitcairn, where they decided to settle. The *Bounty* was burned to avoid detection by passing ships. Meanwhile, HMS *Pandora* was sent out from England to apprehend the mutineers. The mutineers on Tahiti were found and returned to England for trial. By the time the mutineers on Pitcairn were discovered in 1808, only one (John Adams) remained alive, and the other thirty-four inhabitants of Pitcairn were the wives and children of the mutineers, including Thursday October Christian, Fletcher's son. One of the most authoritative and readable books on this infamous mutiny is Caroline Alexander's 2003 book, *The Bounty*.

Our arrival on Pitcairn was anticipated. Chris had visited years earlier on the *Yankee* and had communicated with Pitcairn before this call. Because no safe anchorage exists on the island, a rotating skeleton crew of seven remained aboard the *Verona* while the others were hosted by Pitcairn families. The experience of living with and breaking bread with these resourceful families was unique and inspiring. I found these geographically isolated folks to be remarkably savvy and curious. With good radio communications, a substantial history of interaction with New Zealand and Australia, and occasional passenger ships stopping briefly for them to board and sell their wood carvings, the islanders possessed extensive knowledge of global events and distant lands. In fact, Betty Christian, the wife of radio operator Tom Christian, was soon going to Fort Lauderdale, Florida, for studies.

Bringing a longboat safely into wave-pounded Bounty Bay is a challenge that these seafaring people had mastered over the generations. Most of Pitcairn's children, along with Australian-born pastor Walter Ferris and his wife were assembled at the waterfront to greet the *Verona*'s visitors. As we walked up the steep hill to Adamstown, the children enjoyed the festive moment and their holiday from school. At the top of the hill, we received warm greetings from the older men and women of the community. Each crew member was "adopted" by a family; I lodged with gracious Pastor Ferris and his wife. Although the day had been long and exhausting, after dinner we all gathered at the community meeting hall for introductions and a movie.

My first full day on Pitcairn was spent exploring Adamstown and meeting some of the eighty or so islanders. Early favorites included Vincent and Cairn Christian, bachelor brothers whose home was next door to Pastor Ferris's. As I sit now at my computer, I glance up at a handsome carved turtle created expertly from exquisite hardwood. On the belly are the words "Made by Cairn Christian, Pitcairn Island," reminding me that I had mailed it to my parents in San Antonio as a gift. I suspect that they, especially my dad, had proudly shown it to friends and visitors.

After purchasing Pitcairn Island stamps at the post office for collector friends at home, my second day began at the medical dispensary, where I found a rather long line of patients assembled. Some of them had never seen a doctor, and as a young physician, my sense of uncertainty blended with unmerited confidence that they would benefit from their encounter with me. Three of my patients had minor infections, which required common, available medications. Three others, all elderly, suffered from serious chronic noninfectious illnesses (skin cancer, stroke, and hypertension) for which I could offer little beyond reassuring words and modest advice. The seventh patient, John Christian, a de facto leader and island elder, presented the greatest challenge. He had returned several months earlier from New Zealand, where he had undergone cataract surgery for his left eye. Additionally, eye surgeons in Auckland had performed another procedure on his right eye, and this intervention had not been successful in saving vision. The operation had left him in recent times with maddening pain in his right eye, which interfered with sleep and grossly detracted from his quality of life. Knowing virtually nothing about the surgical specialty of ophthalmology, I could not imagine being helpful, but nevertheless I carefully examined his eye and to my surprise detected a tiny trace of what appeared to be foreign material at the outer side of his eyeball. With forceps, I gently pulled on it, slowly realizing that it was a piece of plastic tubing. Uncertain of what it was, and why it was there, I described it to John Christian and asked if he would like me to remove it. His response affirmative, I cautiously pulled on the tubing until I had removed it from the eye socket. His relief was immediate, and so was mine. John Christian was so appreciative that he rewarded me with a hug and profuse thanks.

The word about John Christian must have circulated quickly because suddenly it seemed that my status as a caregiver had been elevated. I was taken to see Thursday October Christian's house and the grave of John Adams, the only surviving mutineer when Pitcairn was "rediscovered" in 1808 by the American seal boat the *Topaz*. Shortly afterward, I was invited to have lunch with Len and Thelma Brown. Thelma was Tom Christian's sister and daughter of Flora Christian, the elderly lady I had attended that morning (with hypertension and a recent stroke). During lunch, I learned that Len and Thelma had hoped I would be assigned to stay with them instead of with Pastor Ferris.

Lunch with the Browns was memorable, meriting a detailed description in my logbook ("Superb meal with the Browns—fish [dream fish because it makes you dream], chicken, potatoes, soup, pumpkin, salad, bread, ice cream, Jell-O, fruit pudding, and coffee. Fantastic!") That afternoon, I received the gift of a knife from "Auntie Edna," whom I had attended to that morning. Then I returned to the *Verona* for two days of skeleton-crew duty, trying to piece together the cascade of events that occurred during my first days on the island.

Returning to shore two days later, I had dinner at Pastor Ferris's. I learned from him that the best carver of canes, fish, and birds was Len Brown; of turtles, Cairn Christian; and of wheelbarrows, Jacob Warren. My meticulously crafted Jacob Warren miniature wheelbarrow now sits next to Cairn's turtle on my bookcase. After attending to more medical issues the following day, I hiked with Cairn and Purvis Young to the far end of the island to see five-hundred-year-old rock paintings. Then I returned to the Browns' home for another spectacular meal (salad, five vegetables, three meats, four desserts).

My final day on Pitcairn, August 6, was filled with medical consultations, making a tape recording of islanders, taking photos of

John, Vince and Cairn Christian, receiving gifts, and mentally preparing myself to once again part with newfound exceptional friends. During my final visit with Len and Thelma Brown, I was presented with a gift that is the most deeply meaningful item I acquired during the year: a nail from the *Bounty*. I learned that after the *Bounty* was burned and sank on January 23, 1790, an effort was subsequently made to salvage nails from the wreckage. These prized artifacts are treated with reverence by Pitcairn islanders. My treasured *Bounty* nail hangs nearby in my office, framed with the original handwritten note, which reads, "Pitcairn Island, August 6, 1965, To 'Doc' Barney with the compliments of Len and Thelma Brown, 'Bounty Nail.'"

Most of Pitcairn's young men accompanied us in the crowded longboat that carried us on our last trip from Bounty Bay to the *Verona*, where farewells were again exchanged. At 1800, the *Verona* set sail for the Marquesas Islands as we listened to farewell songs from the longboat, and Pitcairn slowly disappeared into the sunset.

Figure 11. Longboat farewells as the *Verona* departs Pitcairn Island. Photo by author.

August 16, 1965

Log entry: *"Low morale among crew as discontent and personal conflicts mount. Partly to blame is long passage and storms with heavy seas. Reach Nuku Hiva late and under full sail speed into harbor. Captain under the influence of alcohol."*

August 17, 1965

Log entry: *"Arrive 01:00. Anchor for night. Officials board 09:00; Chris goes to shore (drunk) with gendarmes at 10:00. Returned 15:00 somewhat more sober."*

August 20, 1965

Log entry: *"To Anaho Bay early. Arrive 10:00. Work day cleaning ship. Not cleared by officials to go ashore."*

September 4, 1965

Log entry: *"Arrive Fare, Huahine Island, later in day anchor in bay near Haapu village...Overcast and rain. 07:00 swim to shore with clothes in bundle overhead. Go to local store, dry, dress, eat bread and visit with children. Later, a young intelligent looking boy runs off and returns in a few minutes with his father, who asks me to follow him home. In large, kerosene-lamp lighted room met wife, daughter; chat in pidgin French; served supper of eggs, coffee and bread. Learn few Tahitian words: maytay (good), roa (very), mai-ti (wind), yorarana (hello), ua (rain), papee (water) and fare (house). Learn that host Francois is village pastor.*

Invited to spend the night; pastor insists that I sleep in his bed despite my protests."

September 5, 1965

Log entry: *"Arise 06:00, breakfast at 07:30 with pastor and family. See rest of house: large, clean and modern, immaculate throughout. All barefoot on "tile" floor. 08:00 George (son, age 11) takes me to Verona in outrigger canoe, show him around and give him gum, return to shore to visit with new friends, and later George takes me back to Verona. Return in evening, walk with Francois, George and other kids to bridge connecting Huahine Niu (large) to Iti (small); lovely walk and scenery, terrific kids. Dinner on board, poker game, to bed early."*

September 6, 1965

Log entry: *"Labor Day in U.S. Awake 06:00, listen to world news (Pakistan-India war, hurricane Betsy hits Florida, steel strike averted, etc.) Bird hunting with Pastor and George, good hiking through jungle but nothing shot. Invited for "tea" at Pastor's 16:00, a feast consisting of roast pork, chicken, taro, breadfruit, coffee and cake. Family taught me that to display pleasure with meals in their culture it is important to make loud noises, slurping, belching, etc., the more the better."*

September 7, 1965

Log entry: *"Skin-diving with Lloyd, Edgar, Hank, Kip; saw large group of giant rays swimming in formation, and several sharks. Return Verona in afternoon and receive visit from swarm of school* *children. Distribute gum (popular), show picture books and discuss geography with globe. Return to shore 18:00 for roast pork dinner."*

September 8, 1965

Log entry: *"Lazy day aboard, cool, poor weather. Afternoon go ashore, invite Pastor and family to Verona, visit, take photos, make tape recording, exchange gifts (beautiful shell necklace). Another difficult farewell. Midnight: anchor dragged, Verona rested on shoals, no damage."*

September 9–10, 1965

Log entry: *"Depart Huahine 09:00; and enter Teanaroa Bay via Toahotu Pass and arrive Fa'a'aha, Tahaa Island about 13:00."*

Tahaa, roundish and indented by several bays, is about five miles in diameter. Immediately to its south is the somewhat larger Raïatáa. Both islands are surrounded by an almost contiguous coral reef with narrow, potentially treacherous passes that offer access to the islands. Perched on the first spreader of the mainmast, I was treated to a bird's-eye view of the reef and pass and approach into the bay.

Weeks earlier, the *Verona* had spent nearly a week at Tahiti, having arrived at Papeete, the port city and governmental center for French Polynesia. My log notes help me recall a blur of activity that included visiting popular drinking holes, falling in with a group of French legionnaires (from Spain, Poland, Germany, Hungary, Vietnam, Italy, and China), circling the island on a motor scooter, visiting hospitals, attending a Rotary meeting as a guest, finding a ham radio operator to send messages home, shopping, eating some nice meals, and meeting quite a few Americans who had arrived in Papeete by sailboat and jumped ship and refused to continue on the same boat for a remarkable variety of reasons. Skippers of "stranded" sailboats were desperately searching for replacement crew members, offering a diverse menu of destinations for young, adventuresome sailors who had minimal aversion to risk. Because Air France Boeing 707s were now arriving on a regular basis, and thousands of legionnaires were guarding France's "secret" nuclear weapon project on nearby Mururoa, Papeete had ceased to be an isolated, exotic destination. As much as bustling Papeete was entertaining but disappointing, the astoundingly open and hospitable culture I later experienced on Huahine tugged at my deep emotions.

As we left Papeete, our westward route had taken the *Verona* to remote parts of French Polynesia where the local culture was remarkably intact despite centuries of Western influence. In Haapu, I had been virtually "adopted" by Pastor Francois and his family, and on the island of Tahaa, the experience was more or less repeated. It seemed to me, Polynesians treated young foreigners as we Americans would respond to a cute puppy. Given the opportunity, we would take the puppy home to provide food, warmth, and affection until it became necessary to return it to its owner.

Now at the island of Tahaa, anchored near the village of Fa'a'aha, several of us jumped into our launch to explore the area. We decided to find Haamene, where an acquaintance had spent time, and stopped at a home halfway into the bay to ask about a trail. There, we met a father and his son Ramon (about seventeen), who accompanied us on the trail to Haamene. We found a store and met its kind Chinese owner, who gave us cakes, saying, "They are one day old and should be eaten."

After purchasing some needed items at the store and returning to Ramon's house, his father gave him permission to spend the night with us on the *Verona*. A few of us spent the next morning spear fishing and collecting shells with Ramon as our expert guide, and in the afternoon, Ramon asked me to walk with him to his house. We arrived late afternoon at their simple but immaculate home on the bay and waited for his parents and sister and other relatives to return. When they arrived, they carefully washed their feet before entering the house barefoot. I was distracted by the beauty and grace of Ramon's sister, Fifi, age fifteen, and I detected a degree of parental concern as they must have sensed hormones at work. Younger siblings Johnny and Violet seemed quite comfortable with a foreign visitor in the house. Later, about dusk, "coffee" was served in the simple dining room built apart from their house, and again feet were washed before entering the dining room. Traditionally, they explained, the larger meal is served midday and "coffee," served with bread and sugar and coconut milk, comprises the evening meal. A blessing was said before the meal, and having learned from Pastor Francois Teio on Huahine about expressing gustatory satisfaction, I joined the others at the table by eating with gusto and ample slurps and other sounds. Their facial expressions conveyed that they noted and approved.

Neat and clean like everything else, the house had some western beds as well as mats, a chest with mirror, an ancient sewing machine, and a Tahitian-language bible published in England in 1881. Also, there were a few books, school pads, and Coleman and kerosene lanterns, which provided adequate light. The evening for me was somehow magical, ethereal. Surrounded by natural beauty and by a graceful and openhearted family, we joked, looked at photos of family members (two older brothers lived in Papeete); Ramon played his ukulele; Fifi danced; we exchanged the English, French, and Tahitian words for days and numbers; and we shared addresses. Sensing that I had been "accepted" by the family, and as a full moon rose over the coconut palms, I experienced profound contentment. I confess that the disquieting (and naive) thought of discarding my world and returning to this one played in my delirious brain as I drifted off to sleep, with the family in their one big room.

September 12–15, 1965

Log entry: *"Drop anchor off Bora Bora Hotel about 14:00 after passing French warship at pier."*

Bora-Bora, just northwest of Tahaa Island, is a place of dazzling beauty. Approaching Bora-Bora, one sees the jagged peaks of Mount Otemanu, once an active volcano, towering near the center of the island. Blessed with a wonderfully protected lagoon, surrounded by reefs and sandy islets, and reachable by boat only through a narrow passage, it is a vacationer's and skin diver's paradise. My Bora-Bora memories are a blur of lighthearted pleasure, spending time with welcoming local inhabitants and with other young travelers, several of whom were invited to spend time with us on the *Verona*. The generosity of Polynesian youth was almost embarrassing, because for them the concept of personal "wealth" did not exist. If they had money in their pockets, it seemed their "obligation" to use it to entertain or feed their companions; mere possession of cash did not seem to have innate value. Of course, it was again painful to depart from Bora-Bora, leaving behind a cluster of loveable new acquaintances.

September 19, 1965

Log entry: *"Arrived Palmerston about noon. Men board Verona, led us through reef, and go to shore with them in motor launch."*

In addition to the "mutiny" prevented the following day, for me, Palmerston stands out for other reasons, which my log, memory, and photographs permit me to recapture in detail.

First is Palmerston's unique history, which was a factor in the warm reception extended to the *Verona*'s crew. After the island was discovered by Captain Cook in 1774, Englishman William Marsters reached this tiny atoll (with a land mass of about one square mile) in 1863 with a Polynesian wife. He later added two more wives. He sired twenty-three children by some reports, and the three families we encountered on the island were descended from Marsters and his three wives. Each family actively competed for our attention as we were sequentially invited to partake of feasts at their respective homes. Three days were largely spent moving from one feast to the next.

The second reason Palmerston stands out in my memory is Palmerston's extreme isolation. A December 30, 2013, article by Thomas Martienssen for BBC News begins: "It is one of the most isolated island communities in the world. The tiny Pacific island of Palmerston is visited by a supply ship twice a year—at most—and the long and hazardous journey deters all but the most intrepid visitors." Palmerston is too far to reach by helicopter, too small for a plane to land, and the surf too rough for a seaplane to surface. For an inhabitant to see a Cook Island doctor requires sailing about 310 miles to Rarotonga. Consequently, my medical services were in demand after the word got around that the *Verona* carried a physician. Between medical consultations, feasting, roaming the tiny island, and skin diving in the atoll's fabulously clear lagoon, I learned from Bob Marsters that William Marsters's three wives were Stepou (from Rarotonga), Tata (from Penrhyn) and Matavai, who had been abandoned on Palmerston by her former husband, John Hernandez, a friend of William Marsters. I also learned that a significant number of islanders had migrated to Rarotonga for their education. Some of them ended up in Auckland, New Zealand, where opportunities for advancement were greater still.

A third reason Palmerston stands out is my memory of a dance held for visitors and islanders the evening of our second day. The venue was a church constructed of timbers from a French cargo ship that had wrecked there around 1913. The social event was spontaneous and joyful. One mental image stands out in my memory: that of our young crewmate Sam Reese Sheppard beaming and smiling for the first time since we left Panama.

Figure 12. Sam Reese Sheppard, at left, Palmerston Atoll, Cook Islands. Photo by author.

Readers who belong to my age group will recall that in 1954, in Cleveland, Ohio, Sam's father, Dr. Samuel Holmes "Sam" Sheppard, was accused of murdering his pregnant wife, Marilyn Reese Sheppard (Sam Reese Sheppard's mother), as the seven-year-old son slept in an adjacent room. The crime ranks among the most notorious and intriguing murder cases of twentieth-century America. After being convicted of murder and spending ten years in prison, Sheppard was freed pending a retrial, at which, represented by famous attorney F. Lee Bailey, his previous conviction was overturned. Numerous books, movies, and a long-running television series were inspired by this case, often depicting a wrongly accused doctor escaping custody and eventually searching for and finding the true murderer.

Whatever the truth may be, at age seven, Sam Reese Sheppard's life became a nightmare. His mother was viciously murdered, and his father was sent to prison. Ten years later, as a *Verona* crew member, he was spared seeing the frequent newspaper headlines about his father as the September 1966 retrial approached. While the identity of our crewmate was common knowledge, the subject was never discussed, and Sam, quiet, well-mannered, and serious, was not treated differently from anyone else. To my eyes, however, a smile never appeared on his face before that dance on Palmerston.

Finally, the fourth reason Palmerston stands out in my memory was something that happened decades later, when I served on a World Health Organization (WHO) committee responsible for oversight of a global program to eliminate lymphatic filariasis, a disabling and often grotesque parasitic disease believed to affect more than one hundred million people living in the tropics. Our committee, known as the Global Program Review Group, met periodically in Geneva, Switzerland. One of the committee members, representing the nation of the Cook Islands, was Dr. Joe Williams. Joe is generously blessed with grace and charm and a gift for politics (he served briefly as the prime minister of the Cook Islands). Joe and I struck up a conversation after the first day's meeting and decided to meet for a drink and dinner that evening. As we got to know each other, I mentioned my South Pacific travels on the *Verona* and my memorable visit to Palmerston some thirty-five years previously. Joe's eyes widened as he exclaimed excitedly, "Palmerston is my ancestral home!" A direct descendent of William Marsters, Joe was astounded to learn that I, but not he, had been there, and he expressed the hope that one day I would accompany him there for a visit. That quixotic visit has yet to occur, but Nancy and I both had the distinct pleasure of spending time with Joe and his lovely wife, Jill, on Rarotonga, where they have a second home. The occasion was a meeting in Rarotonga of the Pacific Programme to Eliminate Lymphatic Filariasis (PacELF). At this time, Joe's "day job" was his prominent and successful medical practice in Auckland, New Zealand.

Figure 13. Author and Nancy, Jill and Joe Williams, Raratonga, Cook Islands, 2002. Photographer unknown.

September 22, 1965

Log entry: *"Palmerston to Apia, Samoa passage day #1. 11:00 crew was asked to meet on deck. With courage, Chris offered sincere apology which was well received by all. Promised not to drink (alcohol) until Hong Kong."*

October 5, 1965

Log entry: *"Crossed international date line; day lost!"*

October 6, 1965

Log entry: *"Arrive Suva, Fiji Islands 09:00."*

October 7, 1965

Log entry: "*Article in Suva newspaper about arrival of Verona and a note about Stellan (Swedish Count).*"

October 12, 1965

Log entry: *"Anchored off Mbengha Island for fire-walking ceremony; new cook John Narayan prepares first meal, very good."*

October 20, 1965

Log entry: *"Arrived Lenekal Bay, Tanna Island, New Hebrides late afternoon. Bob Paul, previous acquaintance of Chris, boards. Drop anchor, Chris to shore overnight and crew unhappy to be 'stranded.'"*

The eighty-three islands of the New Hebrides and the island of Tanna, in particular, merit special comment. The New Hebrides were populated primarily by Melanesians and were governed jointly by Britain and France as a "British-French condominium," which began in 1906 and ended with the establishment of the independent Republic of Vanuatu in 1980. This unusual dual governance evolved because the islands in the late eighteen hundreds and the early nineteen hundreds had attracted many settlers, mostly farmers, from the UK and France. Almost totally duplicated civil services were established, including post offices, but the court systems were merged.

On shore in Tanna the morning of October 21, most of the crew decided to walk to the six-hundred-foot-high active volcano, said to be the most accessible one in the Southern Hemisphere. I chose instead to walk toward Bob Paul's plantation and trading store but happily was picked up by Guy Wellington, the British district agent for Tanna. Pleased to see a newcomer, Guy was cordial and invited me to accompany him in his jeep on his official "rounds." The first stop after Bob Paul's store was the medical dispensary, where I met Fijian Dr. Mua and a missionary nurse. The second was the office of the French agricultural agent. Guy then met with natives in a village to settle a legal problem stemming from a pig that had damaged a neighbor's garden. The neighbor demanded monetary compensation. Although it was not clear to me how the dispute had been resolved, I was confident that I had just witnessed the application of Solomonic wisdom by a Brit from Gloucestershire in deep western South Pacific!

Soon I was sharing lunch with Guy and his wife, Susan, and their three children in a modest, welcoming home. The afternoon passed quickly with a tour of the Presbyterian Mission Hospital, meeting various missionary "sisters" and the pastor, and being introduced to Keith, a British Volunteer Service Organization (VSO) worker, analogous to American Peace Corps volunteers. Amazingly, we both knew a VSO volunteer whom I had met during my Peace Corps service. All these folks expressed a desire to visit the *Verona*, and afterward, at sunset, I returned to Guy's house for a glorious hot bath and scotch before everyone gathered for a feast at Bob Paul's home. And a feast it was: an array of alcoholic beverages plus lobster curry, barbequed goat, fruit salad, and cheese cake. The eventful day was topped off with movies of the *Yankee*'s visit there years earlier. I slept soundly that night, at Dr. Mua's home at the hospital.

As enjoyable and action-packed as this day was, the most intriguing memory and story associated with Tanna stems from a man called John Frum. Some believe that John Frum was a black American serviceman ("John from America") handling cargo during the World War II American buildup toward the Japanese mainland and that he inspired the Melanesian population's leap into a magical explanation. But other evidence suggests that "John Frum" existed earlier. Whether flesh and blood or the stuff of legend, John Frum is part of the deeply fascinating, complex anthropological phenomenon "cargo cult" apparently born on Tanna. The following description is from a February 2006 article by Paul Raffaele in the *Smithsonian Magazine*:

> This is February 15, John Frum Day, on the remote island of Tanna in the South Pacific nation of Vanuatu. On this holiest of days, devotees have descended on the village of Lamakara from all over the island to honor a ghostly American messiah, John Frum. "John promised he'll bring planeloads and shiploads of cargo to us from America if we pray to him," a village elder tells me as he salutes the Stars and Stripes. "Radios, TVs, trucks, boats, watches, iceboxes, medicine, Coca-Cola and many other wonderful things.
>
> The island's John Frum movement is a classic example of what anthropologists have called a "cargo cult"—many of which sprang up in villages in the South Pacific during

> World War II, when hundreds of thousands of American troops poured into the islands from the skies and seas. As anthropologist Kirk Huffman, who spent 17 years in Vanuatu, explains: "You get cargo cults when the outside world, with all its material wealth, suddenly descends on remote, indigenous tribes." The locals don't know where the foreigners' endless supplies come from and so suspect they were summoned by magic, sent from the spirit world. To entice the Americans back after the war, islanders throughout the region constructed piers and carved airstrips from their fields. They prayed for ships and planes to once again come out of nowhere, bearing all kinds of treasures: jeeps and washing machines, radios and motorcycles, canned meat and candy.[1]

According to other Tanna expat residents, I was told that the John Frum movement was dead but had been temporarily revived by the *Yankee*'s visit in the mid-1950s. Several of us were taken to a mountaintop to witness a remarkable engineering feat: a landing strip able to accommodate a DC-3. It seems that this particular cult leader had succeeded in convincing his flock that the reason for American goods not arriving was because they would arrive by aircraft, not in a large white ship. After laboring and completing the landing strip, they sat on it for months waiting for the plane.

I also learned that in an effort to stifle the extensive cargo-cult movement that seized much of the Melanesian world during World War II, American and Australian authorities dreamed up a solution. They would send delegations of indigenous leaders to visit factories that manufactured trucks, planes, motorcycles, washing

---

1. Paul Raffaele, "In John They Trust," *Smithsonian Magazine*, February 2006, https://www.smithsonianmag.com/history/in-john-they-trust-109294882/.

machines, and the like. Observing the manufacturing process, they reasoned, would demystify these items and make it clear that they were not created by "magic." The effort was a dismal failure because the indigenous leaders, after visiting the factories, asked their hosts for the magic to create the factories.

Curious readers who search for more information on Tanna and John Frum and cargo cults may be surprised by the copious amount of popular and scholarly information they encounter. I even read about another group on Tanna who now considers the Duke of Edinburgh (Britain's Prince Philip) to be divine.

October 22, 1965

Log entry: *"Breakfast with Mua, discuss America, etc. Very alert, witty, well educated. Wife physical therapist. Spend morning loafing, photographing, visiting with Bob Paul (drink real planter's punch!) Industrious, imaginative & capable man. 12 years on Tanna. Involved in copra, trading store, farming, cattle, airline (sold shares to natives). Is pilot originally from Australia. Wife quite an asset. Lunch with Mua (boiled beef and cabbage), return to Verona for departure. Bob Paul & few locals and Pastor Alex accompany us to Verona. Depart 15:00 for Blackbeach, arrive about 17:00, spend night on Verona."*

October 23, 1965

Log entry: *"Skin-diving in the morning, beautiful coral, clear water. Noon-to shore for "puberty rites." Climb 45 minutes to village; wait until 15:00. Circumcisions (two boys) took place 6 weeks ago and afterwards kept in neighboring village by men only. Can't be seen by women. Elaborate exchange of gifts for* *village that hosted the boys: killing of pigs, offerings of taro puddings, cloths, meat and other gifts which are accepted and carried away. Repeated several times. We [Verona crew] are given a goat and pudding. The women are elaborately dressed and faces painted for the ceremony as the boys appear with a sort of penis sheath, wearing bright costumes. They circle the central pile of gifts, chanting and followed by dancing. We return to Verona at sundown, leaving the goat behind. At 19:00 depart for Tasman Bay, Malekula."*

Figure 14. Boys with penis sheaths, post-circumcision celebrations, Vanuatu. Photo by author.

October 25, 1965

Although we saw relatively little of the island of Malekula, it is one of the most unique of today's Vanuatu. Chris had also been there before on the *Yankee*, where he had met Oscar Newman, visited his magnificent plantation near Tasman Bay, and been shown other parts of the island by Oscar. I appreciated the opportunity to meet Oscar, to absorb with keen ears his intriguing tale spinning, and to partake of his generous hospitality. Among other entertainments, Chris and I watched movies taken by the *Yankee's* skipper Irving Johnson of the famous "land-diving ritual" practiced annually by tribal members on the southern part of nearby Pentecost Island. While it was decades later that bungee jumping became a widely known and popular form of thrill seeking, this courage-affirming rite (*ngol* or *nanggol* in their language) is performed by young men and apparently has its origins in ancient legends that involve marital conflict. It is also believed to help assure a good yam harvest, the principal source of nutrition in this part of the world. As part of *ngol*, men leap from wooden towers nearly one hundred feet high with vines wrapped around their ankles. An ideal land dive is completed with only the hair on the top of the head touching the ground. Modern bungee jumpers: try that!

Malekula, covered by dense jungle, is home to some thirty tribes. The fact that each has a different language speaks to the relative isolation among tribes. As you may have guessed, anthropologists love this place. The best-known tribes are the Big Nambas at the north end of the island and the Small Nambas at the south end.

October 28, 1965

Log entry: *"Arrive Espregle Bay, Malekula about noon. Chris goes to shore to arrange for trip to Big Namba village to leave tomorrow morning. He also advises villagers about medical clinic to be held in afternoon. To shore with Chris about 16:00 for medical clinic. About 10 patients seen and some drugs dispensed. Sleep on Verona."*

Figure 15. Author seeing patients, Espregle Bay, Malekula, Vanuatu. Photographer unknown.

October 29, 1965

Log entry: *"Most crew leave on trek to Big Namba country. Hank, Meridith, John, Don, Stellan and I remain. Wash clothes in fresh water stream. Read, write, miscellaneous chores. Troops return about 16:00, dead tired and disappointed as village was deserted. Only a few Big Nambas were seen. Sail at 02:00 for Espiritu Santo."*

October 31, 1965

Log entry: *"Feeling exhausted, aching and heavy muscles: viral myositis? Epidemic among crew, low fever, some headache, lassitude, lasts 3-4 days. Depart 16:00 for Samarai, Papua New Guinea. Trading boat "escorts" us out of harbor, taking pictures of Verona under sail. Begin 08:00 and 20:00 watch. Still feeling weak. Halloween."*

November 8, 1965

Arrive on Samarai, Papua New Guinea.

November 17, 1965

Anchor off Madang, Papua New Guinea.

November 27, 1965

Log entry: *"Depart noon for Kaohsuing, Formosa from Madang wharf. Farewells to friends. That night active volcano visible about 50 miles distant with spectacular eruptions and lava flow."*

November 28, 1965

Log entry: *"Madang to Formosa day #1. Celebrate Thanksgiving with turkey and trimmings."*

November 29, 1965

Log entry: *"Madang to Formosa day #2. Cross equator. Two waterspouts sighted in afternoon. Still doldrums; under power."*

December 11, 1965

Log entry: *"Madang to Formosa day #14. Overflown by US Navy reconnaissance plane (DC-4)."*

December 12, 1965

Log entry: *"Madang to Formosa day #15. Gale force winds with gusts up to 50 MPH. Waves breaking well over deck. Possibility of not being able to beat up the coast to Kaohsuing."*

The Verona's *crew struggled with this South China Sea gale mentioned in the December 12 log entry for about twelve hours. This was by far the most violent weather and demanding conditions we dealt with during the year. While on deck, crew members were required to wear harnesses to prevent being swept overboard. The experience was harrowing in retrospect, but during the worst of it, we were too busy to be frightened.*

Six

The Most Isolated Tribe on Earth

"Are those arrows pointed at us?"

"They can't be," said Alan. "We left all those gifts on the beach last night. They must be here to welcome us, and they're carrying their bows and arrows and spears with them." An added note of reassurance: "And remember how frightened they were of us yesterday."

Figure 16. North Sentinel islander shooting arrow towards me. Photo by author.

Alan and Helen O'Brien joined our crew just weeks earlier, in Penang, Malaysia, on February 6, 1966. The *Verona,* now in the Bay of Bengal, had completed more than half of its planned voyage. Alan had burst upon the scene like an irrepressible schoolboy, and we quickly bonded. While decades older than I, his infectious enthusiasm, boundless energy, curiosity, and innate charm quickly melted generational barriers. I vividly recall conversations with Alan that took place as we sprawled in the hammock-like netting under the bowsprit. The unending crashing of waves just below us was oddly reassuring, and the brisk sea breeze in our faces stimulating. That refuge became our preferred spot for subsequent exchanges out of earshot of others, learning more about each other's lives and future dreams.

While every port of call opened the door to new and unique experiences, life at sea aboard the *Verona* had now settled into a routine. We had moved through the crew's mutinous confrontation at Palmerston Atoll, but that was months earlier. Stellan Moerner, our Swedish "playboy" crew member and my pal and gin rummy partner, had been driven off the *Verona* by the clique of young male crew who successfully schemed to make life miserable for him. So after more than seven months at sea with mostly the same crew, the arrival of someone as new and unique as Alan delighted me. Along with a partner in an electrical appliance business, by his mid-forties Alan had achieved financial security. Somehow, he had also negotiated with his business partner an arrangement that freed him to explore the globe with rare need to return to his home in California. Alan had watched his father, a railroad worker, endure a slow and difficult terminal illness, and he was determined not to go that route. So a space between exhilaration and oblivion was where Alan often chose to be. His loving and supportive wife, Helen, did not share his near–"death wish" obsession, but she often did join him at the beginning and at the end of adventuresome undertakings. Alan worked for years (as a volunteer) with renowned anthropologist L. S. B. Leakey and his team's pioneering work in Kenya's Great Rift Valley on the origins of humankind; and he and Helen were cofounders (in 1968) of the Louis Leakey Foundation. Alan was knowledgeable about, and deeply fascinated by, Leakey's research. He also labored passionately and mightily to save the California condor, a magnificent bird then on the brink of extinction.

Pursuing a purer genre of adventure, Alan had experienced several crossings of the Alps in helium-filled balloons, with Helen there to see him off on each attempt and when possible, to join him in a French farmer's field to toast with champagne upon his safe landing. He spent years planning and finally executing a boating trip down the Omo River in remote Ethiopia, hoping to become the first man to survive this bold challenge, but an Englishman beat him by one week. Helen was there to greet him on his return to Addis Ababa, and they even managed a visit with Emperor Haile Selassie. Years later, Alan got his wish to die while on an adventure. As a favor to a friend, the owner of a highly respected adventure travel company, he went to India for a week-long elephant caravan trip being piloted for commercial offering. I received a phone call from him days before he left for India. His heart clearly was not in this trip—atypical for Alan. After finishing his elephant ride, Alan wanted friends to see the Himalayas through the window of a leased aircraft. Clouds obscured the pilot's view, and the plane flew into a mountainside. I and many of Alan's friends around the world were stunned by his death. But that was many years after our longstanding friendship began aboard the *Verona,* flourished while I was living in California, and generated for Nancy and me decades of very special memories of times spent with Alan and Helen.

The *Verona* left Penang late afternoon on February 5, powering out until winds filled the sails. Shortly after sunset, those on deck were entertained by a dramatic sky accented by a rising full moon. Our next landfall was Port Blair in the Andaman Islands, where we arrived and dropped anchor at 02:00 on February 9, 1966. Although Chris had communicated with authorities in New

Delhi to request permission for the *Verona* to enter the Andaman Islands, the customary arrival of boarding officials did not occur—the first hint of a problem. So at 10:00, Chris motored to shore to seek permission from the port officials. His facial expressions were never easy to read, but from a distance one could discern Chris's displeasure. On his return, he told us simply that the *Verona* was denied access to the Andaman Islands. No foreign vessels were granted entry without written permission from New Delhi, as these islands were a protectorate of India. We were aware of the existence of a large penal colony, a restricted area. Also, it seemed that the authorities wanted to protect the indigenous inhabitants from outside influence. These, supposedly, were the main factors behind the exclusion of foreigners. As we sailed from Port Blair on February 9, to our surprise, Chris announced that our next destination was North Sentinel Island, part of the Andaman group. Of course, none of us had heard of this place, and it seemed odd to ignore the *Verona*'s denial of entry by the officials at Port Blair.

Our only source of information about North Sentinel Island, the *Bay of Bengal Pilot*, eighth edition, 1953, published in London by the Hydrographic Department of the British Admiralty, stated on page 22:

> Caution: shipwrecked mariners are cautioned that, in 1901, the following places were inhabited by Jarawas, all of whom were hostile, and by Onges, some of whom were hostile and killed on sight: southern and western coast of Little Andaman Island and Hut Bay on its eastern coast; North Sentinel Island. In 1932, all the Onges were reported to be friendly. The Jarawas, all of whom will kill on sight, inhabited the area. The Jarawas also occupied North Sentinel Island.[1]

---

1. *The Bay of Bengal Pilot* (London: Hydrographic Department of the British Admiralty, 1953), 22.

Further, on page 316, we found:

> North Sentinel Island lies about 15 miles westward of Tarmugli Island, the channel being deep and clear of dangers in the fairway. The island is 400 feet high, thickly wooded, and has a level ridge which slopes to its very low NW point (latitude 11 degrees 35 minutes North by 92 degrees 13 minutes East). A coral reef nearly surrounds the island, extending from the coast for distances from ½ to ¾ miles. The lagoons inside the reefs have depths from 3–8 feet. No dangers exist outside the reefs, and there are depths of 100 fathoms 3 miles west of the island. There are several entrances through the reef practicable for boats, and there is an anchorage outside off of them for a vessel in the fine season. Four inlets stand on the reef. Other entrances are on the southern side of the island, eastward of Antinous islet. Anchorage may be obtained at a depth of about 12 fathoms, about half a mile off the latter of the two entrances.
>
> The natives are timid and hostile (see caution on page 22).[2]

We found the British Admiralty warning entertaining but did not pay much attention to it.

The morning of February 11, we anchored outside the reef, perhaps three-quarters of a mile from North Sentinel Island. Exactly as described by the British Admiralty, the island had a fine beach leading to a heavily forested thicket. The water was a glorious blue. With binoculars, we scanned the island and were surprised to see three tiny black men standing on the reef, an outrigger canoe nearby. About six of us, encouraged by Alan, jumped into our launch powered by an outboard motor and

2. Ibid., 316.

headed for shore. Seemingly terrified when they spotted us, the men ran barefooted across the reef to their outrigger and sped toward shore, paddling frantically. They abandoned their boat and almost immediately disappeared into the dense thicket. Because their boat was abandoned on a reef, and we wanted to signal our friendly intentions, we towed it to the shore and pulled it up safely on the beach. Then we spent an hour or two on the beach, swimming and searching around a bit, and returned to the *Verona* at noon for lunch.

Again spurred on by Alan, late afternoon, about eight of us returned to the beach. Ever the amateur anthropologist, he had brought with him items like beads, tobacco, matches, and candy, which he thought would be eagerly received by indigenous people. Assuming the warnings by the British Admiralty were outdated, and reassured by the obvious terror induced by our presence, we did not feel particularly threatened. We split into three small groups, agreeing to explore inland and to meet on the beach in one hour. As we walked perhaps half a mile inland, following small trails to a larger one leading toward the center of the island, we saw an occasional crude lean-to and traces of old fire sites. We heard unfamiliar sounds, and although we did not see anyone, we sensed that we were being observed. When we gathered again at the beach, the sun was low in the sky, and we excitedly shared our observations. Deciding to build a fire on the beach, we seriously considered spending the night to facilitate our encounter with the indigenous inhabitants. But the dinner gong on the *Verona* reminded us that we had no food, so we elected to return to our vessel. After eating dinner, we watched the fire we had left on the beach and were surprised to see it flare up, presumably because wood was being added. We took this as a favorable sign, a welcoming gesture after finding the gifts Alan had left on the beach.

Very early the next morning, before sunrise, Alan, Warren, and I excitedly sped toward shore after we spotted six to eight natives on the beach. We expected a warm welcome because they were not running away as they had the previous day. When some 250 yards from the shore, we counted some fifteen people on the beach, so we pressed onward. Perhaps 150 yards from shore, we took pause because their gestures and yells did not appear welcoming, and incredulous, we realized that the bows and arrows and spears were pointed toward us. Deliberately remaining at a distance out of arrow range, we shut off our outboard motor to observe rather frenzied yelling and threatening gestures. An arrow flew in our direction but fell well short of our launch.

Fortunately, I had camera gear with me and shot off an entire roll of film, mostly with a powerful telescopic lens. Keeping out of arrow range, we then paralleled the beach only to watch them follow. It became distressingly obvious that we were not welcome. Alan insisted that we make one more attempt to leave gifts on the beach in case they had not found those left the previous day. So we raced at full throttle down the beach, zoomed to shore, leaving matches and trinkets. They seemed to hesitate but then followed to collect the items, but their hostile stance did not alter. Dejected, and realizing that to them we were likely "white devils," we returned to the *Verona*. Our consolation: we had decided not to sleep on the beach the previous night. But most profound was the realization that our twentieth-century world still contained "savages" on a remote island that would literally fight off outsiders. We also assumed that we were likely the only outsiders to have ever set foot on this tiny speck of land in the Bay of Bengal.

Months later, I had the film developed, and to my delight, the black-and-white photographs, at least as viewed on the printed contact sheets, were of surprisingly good quality. And even the few Kodachrome slides turned out well.

Figure 17. North Sentinelese preventing our landing. Photo by author.

About four years later, living in Berkeley, California, and recently married, Nancy and I were enjoying dinner and stimulating conversation with friends Fred Dunn and his wife, Daphne, when I related the story of my Bay of Bengal encounter with "savages." Fred, a prominent medical anthropologist (with doctoral degrees in medicine and anthropology), listened intently as his eyes widened. "You have photographs!" he blurted in disbelief. "Yes, only contact prints, but I have the negatives." Fred's incredulous response puzzled me, so I quickly volunteered to drive a mile or so to our home to retrieve them. He then explained that his PhD dissertation focused on the "Negrito" population of the Andaman Islands, a group closely related to the inhabitants of North Sentinel Island, which, to his knowledge, had never been studied. Fred had large prints made from the black-and-white negatives and studied them intensely, examining for example the construction of the baskets they were carrying, measuring such things as the ratio between the length of spears and the height of the natives, and other details understood and appreciated as potential measures of cultural change. All this methodology was new to me, and intriguing. But considering our illegal entry into the Andaman Islands, it was not appropriate to publish either photos or related observations in an anthropology journal.

With the passage of years, the encounter on North Sentinel Island remained a unique adventure to share with friends who were invariably startled by my rare photos. Then, shortly after the December 26, 2004, Indian Ocean tsunami disaster that attracted global attention, I came across a January 4, 2005, Associated Press article by Neelesh Misra that read:

> PORT BLAIR, India—Two days after a tsunami thrashed the island where his ancestors have lived for tens of thousands of years, a lone tribesman stood naked on the beach and looked up at a hovering coast guard helicopter.
>
> He then took out his bow and shot an arrow towards the rescue chopper.
>
> It was a signal the Sentinelese [as they are now known] have sent out to the world for millennia: They want to be left alone" Isolated from the rest of the world, the tribesmen needed to learn nature's sights, sounds and smell to survive.
>
> Government officials and anthropologists believe that ancient knowledge of the movement of wind, sea and birds may have saved the five indigenous tribes...
>
> The Sentinelese are fiercely protective of their coral reef-ringed terrain. They used to shoot arrows at government officials who came ashore and offered gifts of coconuts, fruit and machetes on the beach.[3]

---

3. Neelesh Misra, "Stone Age Cultures Survive Tsunami Waves," NBC News, January 4, 2005, http://www.nbcnews.com/id/6786476/ns/world_news-tsunami_a_year_later/t/stone-age-cultures-survive-tsunami-waves/#.Wg8SgxMfK8U.

Perhaps the absence of welcoming signals in the Andamans had dampened my enthusiasm to learn more about these islands, but layers of my ignorance have been peeled back to reveal a history that astounds as much as it reveals unanticipated mysteries of the world we inhabit. Modern genetic tools have led to a veritable explosion of information about the migratory pathways taken out of Africa by our distant human ancestors. Geneticists from the University of Oslo wrote in the journal *Current Biology* that the Andaman Islanders are "arguably the most enigmatic people on our planet." It appears that by some genetic measures, the Onge and Jarawa people of the Andamans are more "Asian" than "African" and that, isolated for perhaps fifty thousand years, they represent remaining pockets of the Paleolithic migration that led to present-day Asian populations.

The Andaman Islands have long excited the curiosity and imagination of travelers and geographers. In the writing of the great Greek storyteller/geographer Herodotus, the inhabitants of the Andaman Islands were described as cannibals, a description repeated in the tenth century by the Persian navigator Buzurg ibn Shahriyar. Referring to these same islands in the Bay of Bengal, Marco Polo in the late thirteenth century writes about "a most brutish and savage race, having heads, eyes, and teeth like those of dogs. They are very cruel, and kill and eat every foreigner whom they can lay their hands upon." It is rather certain, however, that Marco was repeating the fanciful tales of others. But this image persisted and was even depicted in a fifteenth-century Paris "book of wonder," the medieval equivalent of today's science-fiction literature.

Searching for additional information, I was as giddy as a kid with a newfound treasure when I discovered Adam Goodheart's remarkable article "The Last Island of the Savages." Wrongly assuming that I and my few *Verona* shipmates were among a handful of outsiders to have laid eyes on an inhabitant of North Sentinel

Island, and to have set foot on its shores, I learned of a fascinating series of remote and more recent contacts.

Apparently, the first recorded sighting of the island occurred in 1771, when the *Diligent,* a hydrographic survey ship of the East India Company described "a multitude of lights...upon the shore."[4] No further investigation took place as the *Diligent* sailed on. Nearly a century passed before the Indian merchant vessel *Nineveh* was wrecked on the reef off the island with eighty-six passengers and twenty crewmen aboard. These survivors reached the beach but on the third day were attacked by natives described by the captain as "perfectly naked, with short hair and red painted noses." They were able to fend off the attackers and were later picked up by a rescue vessel, a steamer sent by the Royal Navy. At this time, India (and the Andaman and Nicobar Islands) were part of the British Empire. The British, fond of the practice of using offshore ships and remote islands (like Australia) for penal colonies, had established a prison on Great Andaman island. This prison was later taken over by India and continues to function as a penal colony.

But the rare encounters between North Sentinel Island and the rest of the world, uncovered by Goodheart's research, did not end with the shipwreck of the *Niveveh* in 1867. It seems that a Hindu convict, using a makeshift raft, escaped in 1896 from the main penal colony and drifted some thirty miles to a beach on North Sentinel Island. Not long afterward, a search party found his body—his throat slit, and his body pierced by arrows.

As far as I can determine, the next "outsiders" to visit the island were the few *Verona* shipmates who foolishly ventured forth that sunny day on February 11, 1966. According to Goodheart, subsequent encounters took place eight and nine years later:

4. Adam Goodheart, "The Last Island of the Savages," *American Scholar* 69, no. 4 (Autumn 2000)., 3.

> In the spring of 1974, North Sentinel was visited by a film crew that was shooting a documentary titled Man in Search of Man, along with a few anthropologists, some armed policemen, and a photographer for National Geographic. In the words of one of the scientists, their plan was to "win the natives' friendship by friendly gestures and plenty of gifts." As the team's motorized dinghy made its way through the reefs toward shore, some natives emerged from the woods. The anthropologists made friendly gestures. The Sentinelese responded with a hail of arrows. The dinghy proceeded to a landing-spot out of arrow range, where the policemen, dressed in padded armor, disembarked and laid gifts on the sand: a miniature plastic automobile, some coconuts, a tethered live pig, a child's doll, and some aluminum cookware. Then they returned to the dinghy and waited to observe the natives' reaction to the gifts. The natives' reaction was to fire more arrows, one of which hit the film director in the left thigh. The man who had shot the film director was observed laughing proudly and walking toward the shade of a tree, where he sat down. Other natives were observed spearing the pig and the doll and burying them in the sand. They did, however, take the cookware and the coconuts with evident delight.[5]

Goodheart reported that:

> In 1975, the exiled king of Belgium, on a tour of the Andamans, was brought by local dignitaries for an overnight cruise to the waters off North Sentinel. Mindful of lessons learned the year before, they kept the royal party out of arrow range, approaching just close enough for

5. Ibid.,4.

> a Sentinelese warrior to aim his bow menacingly at the king, who expressed his profound satisfaction with the adventure.[6]

As during my visit in 1966, the exiled king of Belgium fared better than did the film director.

Again quoting Goodheart:

> Unlike the mainland Indians, the black-skinned Andamanese were of Negrito stock, and they lived as hunter-gatherers, subsisting mainly on fruits, tubers, fish, crabs, honey, wild pigs, and the eggs of turtles and seagulls. They were so small as to be almost pygmies: adult males often measured several inches under five feet. The islanders wore no clothing, and few ornaments; neither sex troubled to cover its genitals. (Indeed, Andamanese men often waggled their penises at visitors by way of friendly greeting.) Though not cannibals, they might easily be mistaken as such, for they wore the jawbones of deceased relatives around their necks. Most astonishingly, they had never learned to make fire, counting instead on the occasional lightning strike and then preserving embers carefully in hollowed-out trees. In short, concluded the first official report to Her Majesty's government, "it is impossible to imagine any human beings lower on the scale of civilisation than are the Andaman savages."
>
> As unimpressive as the Andaman savages may have been to the hard-headed colonial administrators, they provided first-class material for the burgeoning field of anthropology. During the first half-century after the British arrived in 1858, a continuous stream of books, reports, and scholarly

6. Ibid.,4.

> articles appeared, often accompanied by handsome photographic plates: silvery rotogravures in which naked tribesmen fished, danced, or brooded picturesquely over pagan talismans. The interesting hypothesis was advanced that they might be the remnant of a primitive race, one that, sometime in the distant past, had inhabited all of southern Asia. Yet suddenly, unaccountably, not long after the turn of the century, the scholarship stopped. The Andamanese, it seems, were no longer considered a fruitful subject of research.[7]

Nevertheless, interest in the natives of the Andaman Islands did not cease. Goodheart writes:

> I came across North Sentinel Island late one night on the other side of the world. Browsing through an online database, I found it mentioned in an article in a small scholarly journal, with an almost offhand reference describing it as the scene of what was probably the last "first friendly encounter" in history. Only in 1991, the author reported, had an Indian government anthropologist, after more than twenty years of unsuccessful attempts, finally managed to interact face to face with the Sentinelese. Intrigued, I searched the Web and found almost nothing: a few sketchy wire-service reports about the grounding of the Primrose, and the home page of an evangelical organization in California that listed the inhabitants of North Sentinel (along with Buddhists, Jews, and "Gays in San Francisco") in its database of 1,573 "unreached peoples."
>
> I remember how unreal the place seemed, that night in my fluorescent-lit office, as I surfed on the oceans of

7. Ibid.,5,6.

information, the island emerging and then submerging again. I think I already knew that I would try to go there.

The anthropologist who had made the first contact, the scholarly journal said, was a man named T. N. Pandit. As I did more research, I discovered that he had also published the only book ever written about North Sentinel: a slim volume, from a small press in Calcutta, titled *The Sentinelese.* I telephoned Pandit in New Delhi, where he now lives in retirement, and he told me a bit about his work in the Andaman Islands.

I was unable to find T. N. Pandit's book, but I did succeed in locating a review of the book. It offered the following:

> Of the little-known Andamanese tribes, the Sentineli are those least known. They have had no friendly contact with non-Sentineli since we know of their existence, i.e. for the past 200 years or so. For all this time (and for an unknown length of time before) they have lived totally isolated lives on their remote island, well protected by stormy seas and dangerous reefs. This book does not so much deal with the Sentineli themselves as with the attempts of Indian anthropologists to contact them. Visits from outsiders are usually greeted with arrow shots and hostility, though more recently gifts of coconuts have been accepted. There are short but fascinating descriptions of a number of brief contacts between 1970 and 1988. The rest of the book describes Sentineli technology which appears to be at the same Paleolithic level as that of all Andamanese tribes, i.e. while they know, use and transport fire, they do not know how to kindle it. That such a society should be able to survive unmolested in today's world is nothing short of miraculous. Indian plans to establish coconut plantations on the

> island have recently been mercifully abandoned so that the Sentineli are, for the moment, safe from computer salesmen and tourists. The book is a slim one, understandable in view of the scarcity of information, but it remains the only one ever published dealing exclusively with the world's strangest surviving people. Like all books in the "Andaman and Nicobar Island tribe" series published by the anthropological Survey of India is small yet well written, well illustrated and well produced.[8]

When I started writing this chapter, I was puzzled about how Chris had decided to stop at North Sentinel Island. Although in Port Blair we had been denied entry into the Andamans, the *Verona* had without hesitation set sail for this tiny and forbidden island. Pondering this mystery led me to the hypothesis that Alan O'Brien had influenced Chris to make this stop. Alan was an avid amateur anthropologist who came prepared with a collection of trinkets for the natives as signs of peace and friendship. It was he that led the charge to land on the island the afternoon of arrival and encouraged us to split into groups to explore for signs of habitation. Alan had for years spent extended periods working with the famous anthropalentologists Louis and Mary Leakey in Ethiopia in their search for early humans. These considerations, and his boundless energy, led me to speculate that he had somehow "negotiated" with Chris a stop at North Sentinel after he and Helen joined the *Verona* crew in Penang. Of course, I had no way to answer this mystery because Chris, Alan and Helen were dead.

I learned later that my speculation was entirely wrong. After my search located Lloyd in Copan, Honduras, and our visit with him in 2012, I contacted Edgar Faust, Lloyd's longtime friend

8. info@andaman.org, "The Only Book Ever Published on the Inaccessible Sentineli," Amazon.com, August 30, 1998, https://www.amazon.com/gp/customer-reviews/R23BKL30PC7ERI/ref=cm_cr_dp_d_rvw_ttl?ie=UTF8&ASIN=8170460816.

and *Verona* shipmate. During a long telephone conversation with Edgar at his home in Knoxville, Tennessee, I learned that it was he and Lloyd that suggested to Chris the stop at North Sentinel! The two of them had observed an Andaman Island patrol boat being repaired in dry dock, making it extremely unlikely that the authorities had the capacity to enforce our denied entry. On this basis, and having read in the British Admiralty that the last report was from 1901, they concluded that the islanders must by 1966 be heavily impacted by outside influences...if not displaying golden arches. Chris readily agreed. That is how we were privileged to have seen and interacted with, albeit from a distance, the most isolated tribe on earth!

I wanted to reassure myself and my readers of the accuracy of the facts I have related in this chapter. So I asked Barry Hewlett, PhD, longtime friend, colleague, and professor of anthropology at Washington State University, if he knew an expert who could confirm the information. Happily, Barry knew the perfect individual, the world's leading authority, an anthropologist he would soon see at an international conference in India. "Also, I saw Vishvajit last week and he said everything in your chapter was correct; no issues" is the message I received.

Seven

The Red Sea Beckons

We had thirteen days of unremarkable sailing on the Arabian Sea from the Maldive Islands to Aden, Yemen, near the entrance to the Red Sea. I spotted porpoises frolicking, watched a distant waterspout that temporarily diverted our course, and pondered the island of Socotra off the coast of Somalia as we neared Aden. The pilot book described Socotra, with about twelve thousand inhabitants, as an unfriendly place that was rarely visited. Crossing the Arabian Sea, the *Verona* was a solitary vessel, but approaching Aden, it became clear that we were competing with freighters, tankers, passenger ships, and other vessels for our respective positions in a busy shipping lane. As Aden itself came into sight, I was startled by the sight of hundreds of vessels anchored offshore. We dropped anchor the morning of March 24, 1966, the flag of our nearest maritime neighbor that of the Soviet Union with hammer and sickle waving in the light breeze.

As our launch delivered us to the harbor, we passed near the Soviet ship *Explorer* and could see its crew focused intently upon us. When we returned hours later, they waved and appeared surprisingly friendly. This long-distance interplay between the *Verona* and Soviet crews continued the following day, perhaps because of our young female crew members. As we returned in the late afternoon, the Soviets mustered their courage and invited us to board their ship. Out of curiosity, we complied. The initial encounter was rather stiff and uncomfortable; after all, we were on opposite sides of a great divide, the long-entrenched Cold War. But gradually, our group of about ten and a similar number of the Soviet sailors began to communicate in halting English. In response to the question, "What does your ship do?" we were informed that it was a research ship. Asking, "What research?" we were taken to an open area on the aft deck with different kinds of paint and were told that their job was to research the durability of paints. Yes, of course, except that the unmistakable feature of this ship was its massive array of electronic gear. It was clearly a spy ship, a floating listening platform, just like ships operated by our navy that are designed to intercept communications of our adversaries.

But electronics were forgotten as young people from two opposing worlds contemplated one another, face-to-face on the Arabian Sea. Suddenly a young sailor with adequate command of English invited us and our shipmates to return at 20:00 to watch a movie. We happily accepted the invitation, and later, about half of our crew decided it would be fun to see Russian cinema. We secured our launch to Soviet ship the *Explorer*, climbed onto the deck, seated ourselves in a row of chairs facing a collapsible screen, and waited for the movie to begin. The projector started to roll, and for a few seconds we saw flickering images on the screen. But suddenly a catastrophic failure of the projector ended the movie and caused a pall to descend over those gathered. Extreme embarrassment engulfed the assembled hosts, and we guests felt their pain. After all, it was only a few years since Soviet technology had elevated Yuri Gagarin to worldwide acclaim by becoming the first human to orbit the earth in a spacecraft, a feat that makes movie projection mere child's play. Frenzied attempts to repair the failed projector lasted for minutes but felt like hours. Suddenly, the deck was clear; our hosts had disappeared. We milled around, discomforted by their chagrin, and wondered if they would ever summon the courage to face us again. After about ten minutes, we started to return to the *Verona* on our launch, with a gray cloud floating above each of our heads, when suddenly the Soviet crew reappeared, vodka bottles in hand, to announce there would be a party in their mess area.

Figure 18. Soviet sailor aboard the *Explorer*, anchored off Aden, Yemen. Photo by author.

We descended two levels of the *Explorer*, where, side by side on benches at a large mess table, our crews intermingled, and soon magic colorless liquid passed our lips and warmed our bodies and souls. Our hosts' ability to speak English seemed to improve, and before long, the scene at the table resembled a gathering of long-lost companions. A box of Yuri Gagarin pins appeared, and each of us had pinned to our shirts the likeness of this handsome young space hero. New bottles of vodka replaced empty ones. We learned that during the nine months that the *Explorer* was away from its home port, crew members had been denied permission to set foot on shore. We soon found ourselves exchanging names and addresses with our new friends, pondering their invitations to visit them in the then-forbidden Crimean port of Sevastopol on the Black Sea, and gratefully receiving a photograph of their "research ship" signed by crew members. Toasting each other repeatedly, I recall one sailor lifting his glass and toasting, "Up with Russian-American friendship; down with the Chinese."

Just as the cabin was starting to spin, and my footing was less than stable, instant silence descended. I looked up to see an officer standing on the stairs that led to the level above. He clearly commanded the respect (or fear) of our non-officer newest best friends. He spoke, saying, "The captain invites the ranking officer and the blonde to visit him in his cabin." I looked around and realized that as the *Verona*'s doctor, I was the "ranking officer." And the blonde was my friend and crewmate, the captain's niece Margaret. The two of us looked at each other, wondering if either of us was up to the duty at hand. But there was no alternative: the officer stood waiting on the stairs for the captain's order to be followed. It was obvious to all that the captain and officers were missing out on the party and that they too longed for distraction from their monotonous jobs.

The captain and a very tall, silent man welcomed Margaret and me into the captain's well-appointed cabin. His accented English was quite understandable as he unlocked and opened a handsome wooden cabinet and from it withdrew a bottle of vodka and a few cans of assorted delicacies. While social graces demanded that we accept the captain's offerings, instinct told me that I must strive to moderate my intake. The tall, silent man remained silent.

Struggling to speak clearly, I responded to the captain's questions about the *Verona* and its crew, and we enjoyed a pleasant if unanticipated visit with the master of a Soviet spy ship. But then the silent man broke his silence with perfect, American-accented English! As we chatted with him, it became obvious that his were the principle "ears" on this floating listening post and that he also may have served as the "political minder" aboard.

The final minutes of the evening were a blur, but a few fleeting images remained. First, our hosts made it clear that we were in no condition to manage our launch. Rather, they would deliver us to the *Verona* in theirs. Second, we invited the crew to visit us the next afternoon because we thought they would be interested to observe up close our handsome vintage sailing vessel. The evening's final memory: our new friends literally carrying us to our bunks because none of us was experienced with the power of that potato-derived elixir that we had consumed in excess.

Morning. Brain fog lifted in painfully small increments. The *Verona* was scheduled to weigh anchor in the late afternoon, reminding us that we had invited the Soviet crew to visit us after lunch. Now, in the cold light of day, this no longer seemed such a good idea. How could we possibly match the peak experiences of the previous evening? And what would they find of interest on a capitalist sailboat? It seemed unlikely that they would be intrigued with our sails and riggings. About twelve Soviet crew members arrived after lunch and boarded *Verona.* They seemed rather listless and bored as we led them around, showing no interest whatsoever in the sailing features of our home at sea. We found ourselves in the large main cabin, which contained bunks and mess tables, the least interesting part of the *Verona,* when one of the Soviets spied a rather tall stack of *Playboy* magazines and asked if he could look at them. Unadulterated madness, chaos, and mayhem reined for a full two to three minutes. In a millisecond, twenty-four hands, bodies attached, flew through the air toward the pile of

Playboys. Center sections were ripped out in an instant and stuffed under shirts, eyes scanning to see who observed the forbidden act. Crouching in corners, under tables, and in bunks, they wildly continued to tear at pages, hiding the treasures with remarkable virtuosity. The frenzy subsided almost as quickly as it began. At this point, and ready to bid us farewell, they jumped into their launch and returned to their spy ship.

About an hour after our visitors departed, the *Verona* headed toward the Red Sea. It was March 27, and we enjoyed a welcomed trailing wind that allowed us, with only square sails set on the foremast, to make good 190 nautical miles on the first day of our Aden-to-Suez passage. We crew members loved it because while on watch we didn't need to work the sails, but rather we could enjoy watching the freighters, tankers, and rare passenger ships glide past us down the main shipping channel in the center of the Red Sea. Day two of the passage brought favorable winds from the south-southeast and 170 miles made good. The next day, the wind direction shifted, requiring us to leave the main shipping channel as we tacked to maintain forward progress. Around noon, several miles off mid-channel, we spied a white ship that had deviated course to follow and eventually circle us. Astounded, we realized as the vessel came close that it was our favorite Soviet spy ship. Through binoculars, we could see broad grins and waves as the entire crew, including officers, were on deck to wave a final farewell to their new capitalist friends! In retrospect, this extraordinary gesture offered a powerful hint of the Cold War thaw that came decades later.

The Aden-to-Suez passage did not get easier. Facing strong headwinds and tacking every four to six hours, demands on the crew accelerated as the relatively carefree previous days dissolved into memory. On March 30, we advanced only seventy-seven miles, the only diversion a large American tanker from New York that detoured to pass near the *Verona* to admire its fine lines and full sails. As the *Verona* slowly progressed up the Red Sea, and the early April breezes became cooler, crew members on watch continued to be taxed by the stiff headwinds from the northwest. Suddenly, and unexpectedly, late morning April 3 (day eight of the Aden-to-Suez passage) came relief. Falling off the wind, Chris ordered the *Verona* to be steered near Hassani Island, about eight miles off the Saudi Arabian coastline. The lee side of this unpopulated rocky island, about two miles long and one mile wide, offered fine protection from the wind. The anchor was dropped, and almost immediately, the avid skin divers grabbed their gear, jumped into the wonderfully clear, blue waters of the Red Sea, and headed for a long, shallow coral reef. The restful break from the routine on open sea was appreciated and enjoyed by all.

At sunrise the next day, I swam to Hassani Island with a fellow crew member, but not finding much of interest, we returned to the *Verona* for breakfast. Afterward, several of us gathered around the *Verona*'s nautical chart and realized that although we could not see it through the morning haze, only eight miles from us was the village of Umm Lajj. Curiosity gripped us. Saudi Arabia had an exotic ring, so without much thinking or planning, nine of us (including three women) decided to take our launch to see the village, expecting to return within a few hours. We told Chris of our plans, and no objections were voiced. We departed about 09:30, but the trip to shore took a bit longer than expected because we passed small fishing boats and detoured to pass near an Arab dhow. Dismay registered in the eyes of every mariner we encountered. No hint of hostility—just disbelief.

Once ashore at Umm Lajj, we were immediately surrounded by a curious, nonthreatening, wide-eyed crowd. Gestures were the initial vehicle of communication as we were invited to follow the villagers to a nearby building. It was the police station we entered, surrounded by the curious entourage. We still did not realize that we, and especially the three women clad in shorts, were to them as strange as aliens from outer space. We were delighted to discover that John Narayan, our Indian/Fijian cook, had learned rudimentary Arabic while working on seagoing vessels. John thus became our interpreter and spokesman. We learned that someone had gone to search for Abdullah, a Jordanian schoolteacher who was fluent in English, but Abdullah was not near at hand. So it fell to John to explain to the apparent top man in the police station that we had simply come to visit their village for an hour or two before returning to our ship. Asked for our passports, we realized that they remained on the *Verona.*

We were then invited to follow the policeman to the principal municipal building in the village. Entering a courtyard through a rather massive wall and gate, we climbed a staircase to the second floor of a large, well-appointed building, where we met Nasser, the second Jordanian schoolteacher. With Nasser's help we again explained who we were and how we got there, after which we were warmly welcomed as honored guests and offered any service we might desire. By now, it was about one in the afternoon, and lunch seemed a reasonable request. Seated on the red carpets, we were soon served bread, olives, jam, and cheese. Fingers were our utensils. The whole experience played out in a calm, soft-spoken, slow-motion surrealistic manner.

After lunch (photo below), Nasser explained that the most important man in Umm Lajj, a prince, was on the annual pilgrimage to Mecca with his wives but that later, the prince's representative would meet with us. Then it was revealed that to welcome and honor the *Verona*'s crew, a feast would be held that evening. One of us (only one) was invited to motor out with a local delegation to extend the invitation to the entire crew. Shortly after the volunteer left, we were told we could do whatever we wished. After looking around the building and compound, as we neared the gate, we noticed two soldiers with rifles guarding the sole entrance (and exit). Outside were throngs of villagers, mostly young people, hoping to catch a glimpse of the aliens. We wanted to stroll around the village but were gently told that was not a good idea because of the crowds outside. Indeed, the implication was that the soldiers were stationed at the gates to protect us from being annoyed by the villagers, who were not permitted to enter the compound. It all made sense; the Saudis were taking good care of their honored guests.

During the afternoon, a few local dignitaries arrived to see us. Communicating in English was initially a challenge, but judging from the visitors' handsome flowing robes, expensive watches, and poised demeanor, it was clear that they were not common villagers. Seemingly enjoying discreet, fleeting glances at the uncovered legs of our three female sailors, they tended to linger as hosts and honored guests sipped tea. During the afternoon, I was fascinated to watch our Saudi hosts pray at the appointed times, seemingly oblivious to the presence of strangers—infidels, at that.

Shortly before sunset, we watched from a second-floor window the arrival of the *Verona* crew. Chris was among them as they disembarked and approached the compound. They had been wise enough to bring their passports. Chris had insisted that a skeleton crew remain aboard the *Verona,* despite repeated demands by the Saudi hosts that *every* crew member attend the feast.

Figure 19. The *Verona*'s crew arriving; lunch is served. Photos by author.

Hours earlier, in preparation for the feast, I watched in fascination as lambs were slaughtered in the traditional manner, adhering to Islamic law. Later, I could smell the fragrance of roasted lamb and herbs blending with the aroma of freshly baking bread. As some twenty *Verona* crewmates and perhaps a similar number of hosts, mostly clad in flowing white robes, gathered in the large room where we had enjoyed our lunch, two vastly different cultures intermingled in a spirit of mutual fascination and respect. The hospitality extended by our hosts was warm and sincere, and extensive conversation erupted despite linguistic barriers. It also became clear that several of our hosts were well educated and had life experiences far beyond the confines of Umm Lajj. Plates laden with food were placed on the red carpets, and the feasting began.

Engrossed in the magic of this extraordinary evening, I had lost track of time when I heard the phrases "You are our honored guests" and "You will spend the night in Umm Lajj" repeated more than once by the prince's representative. Suddenly the room was silent and tense as Chris's angry voice grew loud and confrontational. "We *must* go to our boat tonight. We *must* take advantage of the favorable winds we have at this moment. You *cannot* hold us here against our wills!" Most, if not all, of us crew members were aghast at Chris's lack of civility, and we worried about his escalating anger, his face getting redder as he hurled threat after threat, including retaliation by force by the US military! Gradually, he calmed down as it was again explained that because we entered the kingdom illegally, the authorities in Umm Lajj had no choice but to await instructions from their superiors in Jeddah, the capital. Telegrams had been sent, and we were obliged to wait for the reply that would come soon. Our hosts made every effort to reassure us that we could resume our travels the next day and that they would make us as comfortable as possible. We learned that we would all spend the night at the prince's palace, which was now being readied for us, and that a military truck would take us there shortly. In the morning, we were to return to the same municipal building where we had been hosted.

When the truck arrived, we were escorted by a couple of young, uniformed soldiers and climbed into the open-bed truck for a short drive to the palace. The distinctively un-American term *prince's palace* conjured up visions of luxury. Ushered into a large, somewhat ornate columned room with sufficient single mattresses on the floor to accommodate us, the image of luxury was somewhat downgraded. But the mattresses looked inviting to tired bodies, and it did not take long for each of us to claim a spot. Drinking water was provided, and modest but adequate restroom facilities were available. The soldiers wasted little time taking up their positions near the only door, resting with rifles at their sides, with a distinctly unthreatening demeanor.

Morning arrived on what would have been day number ten of the Aden-to-Suez passage. About 07:00, we were awakened by the soldiers, prone on the floor saying their prayers. Stretching as I awoke and exiting the door of our communal bedroom, I watched the sun rise over the desert, my eyes drawn to a group of heavily laden camels passing nearby. About 08:00, the truck returned us to the municipal compound, where we enjoyed an ample breakfast of bread, olives, jam, and cheese after being told that the telegram had not yet arrived. The assembly of villagers outside the compound had swollen many-fold, a reflection, we began to understand, of our uniqueness in this barren stretch of Red Sea coastline, where large vessels never called and where westerners had never been seen by the vast majority of inhabitants. And certainly not adult women with legs, arms, and faces uncovered! The number of visiting "dignitaries" increased during the morning, creating the appearance of ongoing consultation. The frequency of serving hot tea to the "detainees" and visitors alike seemed to increase in parallel with Chris's growing irritation and frustration.

Somewhat before noon, the prince's representative instructed Chris to move the *Verona* nearer to the village. He refused, citing an adverse wind direction, and again exploded in anger as he had the previous evening, threatening to blow up the town while muttering something about eye for an eye, tooth for a tooth. Finally, he agreed to move the *Verona* but insisted that for safety reasons, he must take four men from his crew with him. Agreement on this negotiating point reached, the captain and others, plus one female crew member, left the compound to return to the *Verona*. Those of us remaining breathed a sigh of relief that the tense and embarrassing confrontation had ended, at least for the moment. After all, we understood the predicament of the local officials, enjoyed the adventure, felt not the least threatened, and appreciated the hospitality offered by our hosts. The afternoon passed quickly in a succession of pleasant activities. I went from window to window and later climbed to the roof for an unobstructed view of the village and harbor. Looking back at me were large crowds of villagers, mostly children, with only the very young girls uncovered. They were handsome children, with glistening white teeth and wide smiles, who were likely celebrating an unscheduled day off from school. Camera in hand, I took many pictures inside the building and had one taken of me sampling a traditional hookah. Later in the day, I finally had an opportunity to engage in an intriguing visit with Nasser and Abdullah, the Jordanian schoolteachers who had been living in Umm Lajj for about a year, and had a picture taken with them. I learned of their impressions of Umm Lajj and of the reaction of the villagers to us. Nasser explained that we were indeed the first westerners they had ever seen and that they had never seen women "free" (uncovered). He said, speaking, I believe, for himself as well for the locals, that we were seen as "supermen," that they envied our way of life, and that they would like to be "brothers" with Americans. I sensed, however, that perhaps we were also viewed as being somewhat less civilized.

Figure 20. Author with Nasser and Abdullah, Jordanian teachers in Umm Lajj. Photographer unknown.

Before sunset, the *Verona* having been moved into the harbor, Chris and other crew members returned bearing requested items such as pipes and tobacco and books. We were told that a telegram had arrived with explicit instructions to treat us with the greatest possible hospitality until a follow-up telegram could be sent, hopefully the following day. An evening meal of lamb, macaroni, bread, and cheese was served, and we returned to our sleeping facilities at the palace. A good cold shower was another highlight of the evening before we retired about 22:00.

The morning of April 6, our day of deliverance, began as we were awakened by the sound of our praying guards. We returned to the municipal building and started to eat our usual breakfast, when we spied the smiling face of the prince's representative. His sweeping gestures indicated that our freedom was imminent. But then his mood suddenly changed as he instructed that the film in our cameras be confiscated. Fortunately, before the soldier took my camera and ripped out the film (eleven exposures lost, mostly of a magnificent sunset and sunrise), I had already stashed away a full roll of thirty-six exposures.

After breakfast, the calm returned, but an air of uncertainty continued to engulf us. We were then invited to be seated as tea was served to everyone. Again, friendly and smiling, the prince's representative sat down, clearly preparing to begin farewell formalities. Before he could begin these symbolically important rituals, however, Chris, oblivious to the context of the moment, abruptly demanded that our passports be handed over.

Ignoring this demand, the prince's representative graciously spoke to us, saying, "I will convey to you the contents of a telegram received this morning from His Royal Majesty King Faisal. 'I warmly welcome you to my kingdom and regret that you were temporarily detained. I hope that you have been comfortable here, according to my instructions that you be treated as my honored guests in the Arabic tradition. I regret that because Umm Lajj is a small town, I was unable to offer you more comforts, and I urge you to return to my kingdom one day and to visit more developed cities like Medina and Jeddah. Finally, I wish you a safe onward voyage.'"

Astounded that King Faisal himself had apparently become directly involved with our Red Sea misadventure and charmed with his gracious words, we replied to the prince's representative that we had enjoyed our visit, and we thanked him for his kindness. We expressed how much we enjoyed our days in Umm Lajj, offered thanks for the hospitality, and said that we would return home with happy memories of our stay here. His reply: "On behalf of the royal family, you are always welcome every day, hour, minute in our country."

Before noon, we were all back on the *Verona,* passports in hand. By noon, the anchor had been weighed, and we were underway. Four days later, after dealing with gale-force winds and high seas for a day, we reached Suez, Egypt, about midnight and dropped anchor.

Unwilling to miss the opportunity, several of us decided to use April 11 for a mad dash to Cairo in a taxi. In those years, car ownership was limited to taxis and the relatively few elite, making it seem that we had Cairo's elegant boulevards and streets to ourselves. Years later, I spent a great deal of time in Cairo on medical-research projects when the traffic jams in this massive city were legendary.

And dash we did…from the National Museum to the Giza pyramids and Sphinx to the famous Shepheard's Hotel roof for drinks to the famous bazaar Khan el-Khalili, all interspersed with gorging on delectable and astoundingly inexpensive food and freshly squeezed juices. This orgy of high life for us mariners was topped off with a nightclub experience where we mixed with sophisticated Cairenes who were exceptionally friendly toward Americans. It was 03:00 before the driver deposited us back in Suez.

Early afternoon of April 12, with an Egyptian pilot aboard, we joined a convoy of ships transiting the Suez Canal. That night, we anchored in Great Bitter Lake and continued with the convoy very early the next the morning. Before noon, we reached Ismailia and took on a new pilot to take us to Port Said. About midway there, perched high aloft, I observed a most unexpected sight: surrounded by nothing but desert was a gigantic sign advertising Johnnie Walker scotch!

We reached Port Said late afternoon, bid farewell to the pilot, and despite threatening weather, we headed north into the Mediterranean Sea for Beirut. We dropped anchor in the Beirut harbor about 10:00 on April 15 and were cleared by the port officials by noon. But I noticed something a bit curious as we reached the Beirut harbor front. Immediately upon setting foot on land, Chris was approached by two Western-appearing men. They spoke for a moment before Chris, with a worried look on his face, climbed with them into a car. Later, I learned about this final chapter of our Umm Lajj misadventure.

Chris related that when the invitation to attend a feast in our honor reached him, he had been appropriately suspicious, especially with the demand that *all* crew members be present. Among the several who remained on board was Kip Durren, the engineer and ham radio operator. Before leaving the *Verona* for the feast, Chris left instructions that if we did not return that evening, Kip should use the radio to contact the *Verona*'s home office, Ocean Academy Ltd. in Rowayton, Connecticut, to alert them to the situation. Responding to this order, Kip worked over his radio for hours until he was able to raise a ham radio operator in Nairobi, Kenya. The Kenyan then conveyed the message to another ham operator who finally succeeded in reaching the contact person in Connecticut by phone, who then communicated with the U.S Department of State in Washington. The Department of State official then contacted the U.S. Embassy in Riyadh, Saudi Arabia, but by now the message had become grossly garbled, indicating that the *Verona* was shipwrecked in Umm Lajj. This led an eager young American Foreign Service officer to jump into a truck laden with blankets and other supplies to rescue us. To the lasting displeasure of those who engineered the dramatic rescue attempt, the Foreign Service officer, having raced all day along a coastline devoid of roads, arrived thirty minutes **after** the *Verona* had departed. Sheepishly, Chris shared the bare essentials of the chastisement he received at the US embassy in Beirut on our arrival. So ended the naive and spontaneous urge to spend an hour or so in a Saudi Arabian village…

Eight

Peripatetic Dreaming II: Formosa to Ceylon

December 13, 1965

Log entry: *"Once in lee of island winds decreased and we proceeded under power to Kaohsiung. Hove to in outer harbor and enter inner harbor at noon. Tied up to Pier #1 at 13:00. Cleared for shore by 15:00. Met Art (3rd mate on Northwestern, a U.S. Flagship with cargo for Vietnam). Have drink with him and Stellan and take in first impressions of China and the East. Partake of average meal with Chris at Garden Hotel. Drink with Stellan, return to Garden Hotel and hot shower."*

December 14, 1965

On watch duty all day, I was able to complete the required paperwork for a train trip the next day to Taipei, the capital of Formosa (now Taiwan). Chris and I enjoyed an excellent dinner of roast duck, shrimp balls, mushrooms, and bamboo shoots at Hoover Hotel.

Extracted from a letter to my parents dated this day:

Kaohsiung is a bustling port town, population about 300,000, so it is no mere fishing village. Our arrival has created quite a bit of interest in the local people, who rarely if ever have seen a sailing vessel like this one. There are constant crowds milling around the pier where we are moored. In fact, as I now look out the window (I am sitting in the deck cabin) I can see many eager Asian faces peering at us with great interest. An article appeared in the local newspaper this morning about Verona (in Chinese, wouldn't do much good to send you a copy), and another group of reporters from other newspapers came around a while ago to collect information. There are about 15 ships in the harbor now, including a U.S. Navy destroyer, and a U.S. Merchant ship on the way to Vietnam with its cargo, so Americans are no novelty in this town, but certainly Verona is. There are also a lot of U.S. military "advisors" in this area.

Tomorrow Chris and I will go to Taipei, the capital, by train. It is about 200 miles to the north, a trip of some six hours. Should be a nice trip—good trains—with excellent service I've been told, complete with lovely Chinese hostesses who bring around hot tea and cool scented towels every so often. A delightful custom, these scented towels. Immediately upon being seated in a restaurant or lounge, one is handed one of these pleasant refreshers to mop the brow and clean the hands. Also they are handed out during the meal at intervals, a heck of a lot better than finger bowls! So last night I had my first *real* Chinese meal, in *China* of all places. Quite good, Cantonese style. But the food in Taipei promises to be much better, including such specialties as Mongolian Barbecue, spicy Szechuan cuisine, etc. I'll report on future gastronomic experiences as they occur. I think there will be many treats in the coming months.

December 15, 1965

Log entry: *"Train to Taipei with Chris departed 09:20. Scented towels, hot tea, nice hostesses, clean and modern air-conditioned cars. Nice scenery. Early lunch in dining car was good. Chris meets an old friend, a missionary and also an Army General who recognized Chris from newspaper article that morning. He translated article. Met other Army officers who recommended Liberty House Hotel. Arrived 15:20 Taipei. Taiwan information service man meets Chris, asks permission to do photo story. Man meets us later at Liberty House. Nice room with twin beds, excellent service including a Taiwanese "houseboy" named Yeh available at all times. Taxi to downtown, walk around market area, very colorful at night, air perfumed with culinary delights. Many sidewalk vendors of foodstuffs. Walk through lobby of Hotel Orient and buy Newsweek and Times. To First Hotel for Mongolian Bar-B-Q: thinly sliced beef, wild boar, mutton, venison, scallions, coriander leaves, cabbage, soy sauce, yellow wine, shrimp sauce, sugar water, lemon water, ginger water, pepper oil, chili, and hot sesame rolls. Colorful and a taste delight. Cooked on grill over charcoal. Return to Liberty House."*

December 16, 1965

Log entry: *"With Chris to U.S. Embassy and USIS office where meet Doug. Photos taken, short story about Verona, contact Captain Tower of U.S. Navy who expected Chris. Suggestions for sightseeing, including trip to Hwalien. To NAMRU-2 [Navy Medical Research Unit #2] where meet USPHS Officer Griffith, brief tour of facilities, meet E. Russell Alexander, M.D. of University of Washington, Seattle. Several friends in common. Interest in Navy's preventive medicine program. Pleasant and interesting visit; requested gamma* *globulin supply [hepatitis prevention] for all Verona crew. Walk to nearby Omei restaurant with Navy folks. Superb meal (shrimp with crisp rice, cucumbers with garlic sauce, diced chicken with pepper oil, etc.) Return to Liberty House, met by Capt. Tower, taken to U.S. Navy headquarters for Chris to plot Hong Kong to Bangkok course to alert the 7th Fleet. Then to China Travel Service to arrange return train tickets to and flight to Hwalien. To National Museum with (with photographer); marvelous displays of tapestries, scrolls, jades, bronze, ceramics. Dinner at Hotel China with photographer and lastly to Champagne Room of President Hotel. Return late to Liberty House."*

December 17, 1965

Log entry: *"Breakfast 06:30, airport by 07:15, flight via CAL (China Air Line) DC-3 to Hualien. 45 minutes. Accompanied by photographer. Rent taxi (400 NT for day); proceed directly to Toroko Gorge—"hand-carved" out of rock mountains—spectacular. Lunch at small hotel at end of Gorge. Meet Candice Bergen and chaperone/photographer (20th Century Fox) Doris Nieh. Join for lunch. Return to Hwalien to performance by Ami tribe of aborigines. Fine dancing and costumes, Polynesian influence?, warm sendoff, much fun but cold front from Mongolia blew in, very cold. To hotel for coffee, return to airport, DC-4 to Taipei (severe turbulence) arriving 17:30. Return to hotel. Talk to Candice on phone."*

Readers who are younger than I may not realize that in 1965, Candice Bergen, age nineteen, was a celebrated fashion model on her way to becoming a film star. She was intelligent, strikingly beautiful, and poised. Her famous ventriloquist father, Edgar Bergen, with his "partner," Charlie McCarthy, was one of the best-known and acclaimed entertainers in the country. She was in Taiwan with a 20th Century Fox film crew filming the movie *The Sand Pebbles*, starring Steve McQueen. Released in 1966, the movie was a box-office success and nominated for many Academy Awards. Having been out of the United States during the preceding years, I was only vaguely aware of Candice Bergen when Chris and I encountered her and her traveling companion, Doris Nieh, in the restaurant. As we were the only four patrons, all Americans, we obviously did not resist the suggestion to join them at their table. A folk-dance program by the Ami tribe afterward had been arranged for Candice, an avid photographer, and she graciously made it clear that Chris and I would be welcome at this (outdoor) program. And later in the day, Candice also invited us to attend a *Sand Pebbles* filming session scheduled the following afternoon in Taipei.

Figure 21. Candice Bergen with Ami dancers in Toroko gorge, near Hualien, Taiwan. Photo by author.

December 18, 1965

Log entry: *"Photography in morning after sleeping late. Return to market area-overcast day-many photos of shop windows with reflection of people in the street. To NAMRU-2 at 11:00, visit with Dr. Alexander and get supply of gamma globulin (courtesy of Navy). Return to Omei for lunch, share table with film studio soundman. To Naval Officer's Club for 25 cent drinks and 10 cent ice cream, fine bar. Arrange for visit to Sand Pebbles set about 18:30 and see street riot scene shot several times. Meet Doris Nieh and fellow photographer Jack and other 20th Century Fox employees. Later to NCO club for "American feast" of milk shake, cheeseburger, pizza. Good night's sleep."*

December 19, 1965

The day was spent shopping, packing, saying farewells, and lastly, attending a performance at the Center Restaurant Theater, which offered good entertainment: superb magician and acrobats and a very good meal. At 22:15, I boarded the train to Kaohsiung, arriving 07:00 on December 20th. Chris had returned earlier that day to Kaohsiung.

December 20, 1965

Log entry: *"Arrive 07:00. Taxi to Verona. Last minute shopping, post cards, etc. Large crowd gathered at dock for 11:00 departure. Administer gamma globulin .05cc/lb to entire crew. Fair winds, 7 knots, smooth seas, should be good passage to Hong Kong."*

Note: No vaccine existed for hepatitis type A at that time. Gamma globulin injections were quite effective in preventing the illness for about six months."

December 22–24, 1965

Entering Hong Kong harbor around 16:00, our entire crew was on deck, excited by the whirl of activity, good weather, the sight of nearby Chinese junks, jets taking off from the airport, and the hint of Christmas in the air. Frank Knight's large and colorful inflatable Santa Claus hung from the second spreader of the mainmast. Health and immigration clearance was quick and efficient after we anchored near the Hong Kong Yacht Club, where we, by prior arrangement, were free to enjoy their facilities—a great convenience. The Yacht Club was filled with friendly folks, mostly Brits, many of them crew members on British Overseas Airways Corporation (BOAC) aircraft. At this time, Hong Kong remained a British colony, long before the 1997 transfer of sovereignty to the People's Republic of China.

During my Bolivian sojourn, I had met Michael Rougier, a photographer sent by *Life Magazine* to cover a grave crisis related to the leftist miners and hostages they had taken, including several American officials and a Peace Corps volunteer. I crossed paths often with Michael during the two frantic days spent near the mines, while officials of the Bolivian government and leaders of the miners thankfully negotiated a way out of the impending civil war. I had been asked by American ambassador Ben S. Stephansky to travel with the negotiating team to the mines to evaluate the health status of the hostages; Mike had been assigned to use his camera to document the impending chaos in remote Bolivia. We kept in touch afterward, and shortly before joining the *Verona,* I spent time with him and his family in South Florida.

Mike was enthusiastic about my *Verona* plans, and when he learned the itinerary included Hong Kong, he insisted that while there I contact his friend Robert Morse, head of the Time-Life office in Hong Kong. Upon meeting Robert Morse, I was delighted to be invited to join him and his wife at their home the following day for Christmas Eve dinner. Among the varied enjoyable experiences of my first full day in bustling and exotic Hong Kong, his invitation was the most welcomed.

American aircraft carrier the *Bonne Richard* was visiting Hong Kong. I met US Navy officer Jim Myers and was invited by him to visit his ship on Christmas Eve day. This seemed a fine time to be surrounded by fellow Americans, and indeed it was. But dinner at Robert Morse's handsome apartment that evening was magical because of the convergence of diverse and fascinating guests, a splendid feast prepared by his lovely Chinese wife, and the stimulating conversation. The war in Vietnam (of which I was virtually ignorant) was in its early stages, and the guests included Time-Life correspondents, photographers based in Saigon, Hong Kong, and Tokyo, and the US State Department official serving as the official US spokesman in Saigon. Near the end of the evening, additional friends arrived: Volker and Hannie Hoffman of German TV, based in Hong Kong. It is hard to conjure up a better introduction to the dynamic world of Hong Kong and to the swirling global events of the day.

Christmas Day, 1965

Log entry: *"Late breakfast with Stellan at hotel. Noon meet Volker and Hannie and their friend German TV cameraman Jan Pieter Bock at Yacht Club and then for "tour" of Verona. Joined by Chris at Yacht Club for drinks and lunch at Diamond Restaurant: great meal, much fun, snake soup, many new dishes, treated by Jan*

Pieter. Then I go with Volker and Hannie to their lovely apartment for coffee, pastries, Galliano and cognac. Listen to music, play with baby daughter, visit their beach club, drinks, return home for nice meal, return to my hotel, write letters, asleep by midnight."

January 4, 1966

The *Verona* set sail for Bangkok late afternoon. My senses were still processing the multiple facets of my Hong Kong sojourn, so I was pleased with the relative solitude at sea. The ten days following Christmas had been filled with newly acquired friends, attending the Beatle's movie *Help!* (my first exposure to the Beatles), sightseeing, a memorable two-day visit to Macao via hydrofoil, photography-related activities, letter writing, and making purchases. Most difficult was bidding farewell to the many fine people I had met.

After the *Verona* set sail at 16:00, I discovered that two new *Verona* crew members had joined us. One was Robert Morris, a pleasant young American who had been in touch with Ocean Academy Ltd. about spending time on the *Verona.* The other was Margaret Gordon, Chris's poised, attractive, nineteen-year-old niece from Charlottesville, Virginia.

January 16–30, 1966: Bangkok, Overland Travel in Thailand, to Cambodia, and to Penang to Meet the *Verona*

Bangkok in 1965 offered much to travelers: a warm, receptive population, splendid cuisine, exotic sights, very reasonable prices, and only modest traffic congestion. It seemed that all Thais were eager to meet and interact with foreigners, especially with Americans arriving on an impressive sailboat. But beyond these welcoming features of Bangkok, the subsequent two weeks provided a succession of rich experiences, none of which could have been anticipated.

Thailand's principal river, on which Bangkok is situated, is the Chao Phraya. Bangkok's most elegant hotels were adjacent to this broad, calm river which empties into the Gulf of Thailand. Because Chris had been there before, he had arranged for the *Verona* to wind its way upriver until dropping anchor directly adjacent to the Oriental Hotel, considered Bangkok's finest. The *Verona*'s crew was offered free access to its swimming pool and related services, including dressing facilities and hot showers with fresh water! Heaven was delivered to our doorstep as we arrived early afternoon on January 16, the twelfth day of our uneventful passage from Hong Kong.

After two days of absorbing the profusion of sensory inputs offered by Bangkok, Ron Buell and Frank Knight and I decided to explore the hinterlands of Thailand on rented motorcycles. By noon on the nineteenth, we were underway, armed with a map and each with a small bag of personal items. Initially thinking that five days would permit us to reach famous Chiang Mai in the north, we soon realized it was not feasible. But we were excited to be out of the busy city and into the countryside, where the flat rice paddies slowly gave way to gentle hills as we entered the town of Takhli late in the afternoon. Deciding this would be a good town to spend the night, and convinced that we were finally far removed from city tourists and all foreigners, we spotted a modest hotel, checked in, showered, and dressed. Soon, with giggling young women knocking on our door and entering, it dawned on us what kind of a "hotel" we had chosen but, regardless, decided to stay the one night.

Later, while standing on the small balcony extending from our room and facing the main road, the young ladies began to speak excitedly to one another. Suddenly and abruptly, they literally pushed us flat to the floor of the balcony and sat on us for a few perplexing minutes. Finally, free to stand, and realizing that the apparent but unknown "threat" had passed, we walked across the road to a simple restaurant to attend to our thirst and empty stomachs. The beer was cold, and the food (spicy beef and rice, soup) was ample. Back on our bikes to check out the rest of town, we were surprised to see three American-looking young men walking by the side of the road. They thought I was joking when I asked if they were Americans. When they heard that we were civilians just traveling through on our bikes, they invited us to visit their air force base and told us how to find it. This was unexpected because an article I had just read in *Time* magazine contained a strong denial by US officials that Americans operated an air base in Thailand in support of the war effort in Vietnam.

Because I had completed two years of service as an officer in a US uniformed service, I carried an identification card issued by the Department of Defense. This card succeeded in getting me past the sentry at the gate of Takhli air base, and surprisingly, I was able to convince the airman to allow my civilian friends to enter as well. Soon, as planned, we found ourselves at the Officer's Club with Lieutenant Flint, whom we had met on the road hours before. As I enjoyed a fifteen-cent Budweiser at the Air Force Officers' Club in a base that did not exist, the combat pilot next to me at the bar was describing vividly to everyone within earshot the mission he had flown earlier that day over Laos. He also described other missions over Vietnam. But our security breech was not long-lasting, because a colonel approached and in no uncertain terms advised us that we didn't belong there. Lieutenant Flint, apologetic, suggested we try the NCO club where, in fact, we devoured a rib-eye steak and trimmings for $1.50 each.

It was late when we returned to our "hotel," where determined effort was required to convince the young ladies who had pushed us to the floor that the only thing on our minds was sleep. And by then, of course, we had figured out that earlier in the day, thinking we were GIs, they had been shielding us from the military police as the police passed an establishment that was clearly off-limits for GIs.

As we continued to meander through the Thai countryside, an experience quite unlike living on the *Verona,* we encountered an abundance of friendly people, tasty meals, village puppet shows, food offerings to a large Buddha with incense perfuming the air, and a wat (temple) teeming with monks in saffron robes. An outgoing young Thai approached us and offered to take us to a Peace Corps volunteer (PCV) named John, who lived nearby. That was like striking gold, because having been around PCVs for the previous two years, I knew that they invariably are the source of priceless local information, far better than a Lonely Planet travel guide hot off the press. PCV John, and later a fellow PCV, convinced us that reaching Chiang Mai was unrealistic and that instead, we should head for Lopburi to see its famous monkey temple and ancient ruins. They also gave us the names of PCVs who lived in Lopburi and suggested a hotel with clean rooms and shower.

A visit to the monkey temple lived up to its billing; I fended off hordes of fearless monkeys trying to grab my camera, and observed their favorite pastime of stealing popsicles from the hands of Thai visitors. These monkeys licked the cooling treat in the universal fashion of a young child. Lopburi became our base of operation for two days as we explored the area, including the large Buddhist temple surrounded by giant bells at Phra Buddha Baht. It was Chinese New Year, and the local custom was to circle the temple with a sturdy stick purchased for a few cents and to bang each bell to insure good fortune during the coming year. Bells still ringing as we left the temple, we were reluctant to part with our cherished "Phra Buddha Baht bell beaters."

Ron and Frank and I returned to Lopburi, planning to leave in the late afternoon of January 22 for Saraburi. But our PCV acquaintances advised us of bandits on the road, stressing that travel after dark was quite dangerous. They explained that the typical technique used by bandits was to obstruct the road with a large tree trunk, stop a bus or car, and rob the passengers. Heeding their advice, we thanked them and quickly headed for Saraburi, which we thought we could reach before dark. But night falls rapidly in the tropics and had arrived before we reached our destination. Suddenly, in our headlights, a massive tree trunk came into view across the deserted road. Feeling terror and making an instantaneous decision, we agreed not to stop but rather to speed up, hold our breaths, and drive around it on the shoulder of the road. We were vaguely aware of figures in the nearby bush as we sped past the log without incident. Before long, we were able to celebrate our "deliverance" with a cold beer at our simple Saraburi hotel. We returned to Bangkok the next afternoon and took special pleasure in swimming in Oriental Hotel's modern pool with the *Verona* anchored in view.

Figure 22. Frank Knight and Ron Buell, rural Thailand. Photo by author.

Examining a map of Thailand and Malaysia, we realized that the *Verona*'s next destination, Penang, an island off the west coast of Malaysia, could be reached far sooner by land than by sea. The former required an overnight train ride plus a final short jump by ferry; the latter required many days of sailing around the long peninsula that constitutes southern Thailand and Malaysia and then, turning north, passing through the Straits of Malacca to reach Penang. Several of us (Lloyd and Edgar, Bob Morris, Margaret Gordon, and I) asked Chris's permission to travel overland and to rejoin the *Verona* in Penang. He had no objections, and we were pleased because this strategy offered us the opportunity to visit famous Angkor Wat.

Periodically, we had met travelers who had just returned from Angkor Wat. They raved about the splendor of these world-famous tenth- or eleventh-century Khmer ruins discovered in dense Cambodian jungle. We were also aware that impending regional warfare made it likely that Angkor Wat might soon become inaccessible. Indeed, a visit to the US consulate in Bangkok revealed that while travel to Cambodia was strongly discouraged, US citizens were free to cross the Thai border into that country. The advice of the young travelers who returned from Cambodia provided the reassurance we needed to proceed to Angkor Wat.

Train tickets and a visa to return to Thailand in our pockets, early the morning of January 25, the five of us were off to Cambodia. In the afternoon, a bus took us from the train terminus to near the border, where we passed Thai immigration and then walked across a narrow no-man's-land to a tiny Cambodian immigration building. The difference from Thailand was immediately obvious: French was the language we heard, the Cambodian people had much darker skin, and poverty was greater. The immigration officials asked us to wait in a nearby bungalow while they prepared our visas, requiring that we leave our passports with them for a short time. Soon, the passports and visas were delivered to us, and we boarded a bus that would take us toward Siem Reap. Changing buses in Sisophon, we reached our destination shortly after dark, found a hotel, showered, enjoyed a surprisingly fine meal, and ended up talking politics until midnight with other Americans we had met on the bus.

Armed with maps and rented motorbikes, the five of us attempted to see as much as possible of the enormous complex of temples surrounding Angkor Wat, truly one of the world's most remarkable treasures. We encountered but a handful of tourists, lost in the vastness of the complex. Then we were surprised to see a lone young woman approach us as we entered Angkor Thom. She identified herself as Ann Cummings, an obviously adventuresome young American, fresh out of college, who decided to spend a year in Tokyo, and who was on her way home traveling across the Pacific. Ann soon became the sixth member of our group and after hearing about the *Verona*, immediately declared her desire to travel with us from Penang to Ceylon. From Ceylon, she planned to fly home. She and Margaret Gordon quickly bonded, Ann the seasoned traveler and experienced adventurer, assumed a mentor's role.

The following morning, we meandered through more of the temple complex. I felt almost overwhelmed by the desire to capture photographic images. The bustling open-air market in Siem Reap abounded in spices and silks, Cambodian art, and other exotic items. Now a group of six, we retraced our steps to Bangkok by train and after a final surge of sightseeing, we boarded a train to Malaysia. This train carried us to the ferry for Georgetown, Penang.

Penang, an island now highly developed and linked to mainland Malaysia by a massive bridge, in 1966 had a slow-paced, welcoming feel. Georgetown, Penang's main city, felt like Hong Kong minus the constant bustle of a few million inhabitants. While ethnicity seemed divided among Chinese, Malays, and East Indians, virtually everyone spoke English. Despite its exotic, tropical character, Penang felt more "Western" than did Thailand or Cambodia.

After settling into a hotel to await the *Verona*'s arrival, we explored Penang on foot and circled the island on rented motorbikes. Good fortune led to a chance meeting in a Penang neighborhood with erudite Mr. G. H. Ho, a retired schoolmaster, and his wife. They offered us much good advice about local sights and happenings and invited us to their home on more than one occasion for refreshments and stimulating conversations.

The *Verona*'s arrival in Penang on February 3 allowed Ann Cummings to successfully petition Chris for her passage to Ceylon. We returned to Mr. Ho's house that evening and learned that for the first time in thirty years, the Chinese, Hindu, and Malay New Years would be celebrated on sequential days. Most memorable, he advised, would be events at a Hindu temple. There we observed a unique rite and procession. Fervent Hindus, fulfilling vows made to their gods, suffered unimaginably from morning until sundown from swords and daggers and other sharp objects penetrating their bodies. Most of the offending weapons were swords supported by a metal platform (*kavadi*) carried on the back and shoulders of the supplicant who, in most cases, was repaying a debt to a specific god. For example, the debt may have been for having saved the life of a sick child or perhaps for having cured a deadly illness. The supplicants were surrounded by family and friends, who offered water and encouragement in the form of singing and beating of drums. The procession of supplicants "carrying the *kavadi*" ended in a frenzy of prayer and pain at the temple at sundown, when the *kavadi* was removed and the wounds attended to. To a Western eye, the scene was off the scale of unthinkable.

Figure 23. Two men carrying the *kavadi*. Photos by author.

February 6, 1966

This was day number 1 of our passage to Fort Blair, Andaman Islands, in the Bay of Bengal. I dealt with crew's medical issues (hives, flu, diarrhea, vaginitis, skin eruptions) that had accumulated. Alan and Helen O'Brien, whom I met briefly the previous day in Georgetown, Penang, were now part of the crew until we reached Madras, India.

February 9, 1966

Log entry: *"Arrive Port Blair; drop anchor 02:00. 10:00 Chris to shore to see officials who refused entrance. 14:00 leave Port Blair."*

February 11 to March 1, 1966

About 19:00, we set sail from North Sentinel Island for Madras, expecting to arrive in about one week. Passing through a broad, lightly traveled portion of the Bay of Bengal, the wind abandoned us the following morning, and Chris gave the order to start the *Verona*'s diesel engine. At 10:00, a fire erupted in the engine room, apparently caused by diesel fuel that leaked onto the engine from a fuel line. Fortunately, the fire was controlled within a minute, but the note in my log book later reminded me of the *Verona*'s fiery ending off the west coast of Africa the following year. Perhaps it began in a similar manner. Around 14:00, a second fire started in the engine room, but was quickly contained. And remarkably, about an hour later a third unrelated fire, this one electrical in origin, started at the port running light. In retrospect, it seems odd that these potentially ominous events failed to spark meaningful concern among the *Verona*'s crew.

We arrived in Madras the morning of February 27. The city spread out before us offered our first exposure to the vibrant, distinct culture of South India, but it was late afternoon before we received clearance to go ashore. Margaret and Ann's friendship, which began in Cambodia, had continued, and the three of us joined forces to explore dynamic Madras. Our first foray was to share farewell drinks with Alan and Helen O'Brien at the Victoria Hotel before they flew to California the next day. The next few days we visited by bus sacred Hindu temples in towns near Madras and tried to absorb the new culture in which we found ourselves. At one of the temples I was fascinated by an ancient holy man, and was able to capture his remarkable image as he prayed to the statue of a Hindu god.

We met young European and American travelers who had explored the hinterlands of India, traveling mostly by train, and several of them encouraged us to visit the heart of South India's

Tamil culture in Madurai, a city some 250 miles to the southwest. Among other attractions, Madurai was said to offer the most spectacular complex of polychrome temples in the entire country. The *Verona*'s scheduled port of call after Madras was Trincomalee, a city on the northeast coast of Ceylon (now Sri Lanka), which once served as a Portuguese outpost after the celebrated voyage of Vasco da Gama opened up a sea route between Europe and Asia. After determining that Chris was willing for us to rejoin the *Verona* in Trincomalee, ever-adventuresome Ann Cummings and I decided to travel together to Madurai before continuing on to Ceylon. Thus, much of the next day was spent obtaining visas from the Ceylonese consulate in Madras and purchasing second-class train tickets to Madurai.

The train left Madras in the late afternoon and was scheduled to arrive in Madurai early the next morning. As we entered our half-full second-class car, we became aware of the intense curiosity we generated as obvious westerners. While selecting our (unreserved) seats, an articulate and confident young man motioned for us to join him. We complied and before long were engrossed in the type of spirited conversation expected among trusted friends. Ernest Hutt explained to us that as an Anglo-Indian by birth, he was trapped between two cultures or civilizations, not truly accepted by either. It was not until reading E. M. Forster's *Passage to India* decades later that I began to appreciate the depth of his dilemma. That night, locked in conversation with this educated young Indian, the hours vanished, and Madurai approached. Ernest Hutt, with his free and easy manner, said to us, "I have a very close friend in Madurai who owes me a favor. You must contact him. His name is Vittal Issac. I will let him know to expect you."

The Taj Hotel near the Madurai train station had been recommended. We went there and enjoyed a hearty breakfast of tomato omelet, coffee, and toast and afterward, on a whim, decided to telephone Vittal Issac. The enthusiastic voice of a total stranger on the other end of the line instructed me to "stay exactly where you are. I am leaving immediately to pick you up." Startled by this turn of events, and not knowing the local culture and mores, Ann (more mature and seasoned than I) put on a "wedding ring" so that we could pretend to be a married couple. We also briefly discussed our "cover story"—i.e., where we lived, how long we had been a couple, and such details.

Thus began one of the strangest vignettes of my young life. It seemed to be borrowed from a farcical Italian opera libretto. We had no idea what to expect, but soon an impressive chauffeur-driven vehicle arrived, and from a rear door popped out Vittal Issac, diminutive, exceedingly dark in complexion, and smartly dressed in white linen. He was warm and outgoing and seemed truly delighted to greet us, and after relating a bit about our travels, he instructed, "Then we must go directly to my bungalow where you can refresh yourselves." While en route, Vittal Issac explained that he was the superintendent of a massive textile company and that his home was located within a large fenced complex. His "bungalow" proved to be a large and elegant colonial-style home, lavishly furnished, staffed with servants. As we sipped tea and tried to absorb this new dreamlike chapter, Vittal Issac went to extreme length to explain that his English wife was currently visiting her family in England, repeatedly emphasizing the fairness of her complexion. He then hastened to show us confirmatory photographs. We had no idea of the degree to which Indians, particularly in South India, were conscious of gradations of skin pigmentation. But I did then recall one of the young foreign travelers in Madras proudly relating his practice of purchasing a third-class train ticket but once aboard moving to a first-class car. When I asked how he could get away with that he replied, "In southern India, no railway employee would ever demand to see the ticket of a (white) westerner."

At Vittal Issac's insistence, tea was followed by a whirl of activities. First, he announced that we must be his overnight guests. After refreshing ourselves a bit, we sipped beer and shared a fine lunch with him and then rested in a comfortable guest room until late afternoon. Then more tea and cakes and off to a Rotary Club meeting, where we were introduced to the membership, which included several doctors. One of the doctors (Dr. Ravi) urged me to remain in Madurai to practice medicine with him. The meeting was over in the early evening, and we were taken sightseeing after briefly visiting Vittal Issac's "Auntie." Afterward, we returned to Vittal Issac's home for whiskey, dinner, and an after-dinner walk to visit the compound's clubhouse. We collapsed into bed about midnight.

The pace did not slacken the following day. Dr. Ravi's driver picked us up and took us to visit the famous Meenakshi Amman Temple, anchoring the massive complex of temples in central Madurai, also known as "temple city." We climbed to the top of the tallest temple, which offered a unique photographic perch.

Figure 24. Temple complex, Madurai, "Temple City", southern India. Photo by author.

We returned to Dr. Ravi's office and then were whisked to his home and introduced to his wife. Soon afterward, we were off to the medical college, where we met the professor of preventive medicine and pathology. Back again in Vittal Issac's home, we were refreshed with cold beer followed by a tasty lunch of fish, chicken, mutton, rice, peas, and more. In the afternoon Vittal Issac took us to see the textile mills he managed. The mills were filled with hundreds of purposeful employees, all of whom appeared deferential to him.

We felt grateful but rather trapped. Aware of the obligatory need to rejoin *Verona* in Trincomalee, we decided to travel that evening by bus to Tanjore and overnight there. The following day we would travel by bus to Tiruchirappalli, from which a Ceylon DC-3 would deliver us to Jaffna and then to Colombo, Ceylon's capital. In this land overflowing with hospitality, I felt a growing and keen discomfort with accepting such generous hospitality from total strangers. It was also obvious that Vittal Issac was rather lonely in his big, empty house and that he thoroughly enjoyed entertaining his foreign guests. He did not hide his disappointment over our decision to leave that evening, but when learning our destination was Colombo, he exclaimed excitedly, "Oh, I have a very close friend in Colombo who owes me a favor. I will send him a telegram immediately to meet you at the airport."

From today's perspective, I felt as though I was being drawn into the subcontinent's version of the movie *Groundhog Day* with Bill Murray. Stunned, but confident that we could exit the Colombo airport incognito, we did not protest his apparent offer to prolong his reign of hospitality. Alas, there was no escape on arrival because waiting for us at the Colombo airport was a smiling couple with their young daughter. They spotted us immediately and whisked us into their car, drove us to their comfortable urban middle-class home, and proceeded to tell us about all the things we would do and see in Ceylon. I believe then we began to wonder if in this part of the world, receiving and entertaining Western visitors conveyed status to the host. Perhaps it was less a question of "a friend owes me a favor" than "I will do a favor for a friend by sending them these foreigners." Just as a head nod in India may convey the opposite of what we understand, the chasm separating American and Indian cultures can lead to gross misinterpretations.

Two days with the Colombo couple proved delightful and offered memorable experiences. The first day was spent sightseeing in Colombo, and included a visit to Colombo's famous zoo. I enjoyed the show featuring trained elephants and their mahouts almost as much as the nine-year-old daughter.

The following day, we drove in the family car to Kandy, a popular resort area in Central Ceylon, with dramatic mountains and lakes and lush vegetation. After sightseeing and lunch, we drove on for a few hours to the village of Sigiriya to see one of the world's most phenomenal sites. The fifth-century Sigiriya Lion Rock fortress was created and once occupied by a ruthless and paranoid tyrant. It was carved into a massive stone mountain rising abruptly from a plain. This storied ancient fortress offered an additional marvel: the nineteen Sigiriya frescos that remain, of several hundred, on cave walls. They are thought to have been painted some fifteen hundred years ago. The origin and interpretation of these sensuous female figures continue to intrigue and puzzle historians and art scholars.

As much as these ancient marvels fascinated us, the need to travel to Trincomalee remained, and ultimately Ann needed to collect her belongings on the *Verona* and travel to Colombo for her flight home. Thanking our gracious hosts, we begged them to leave us in Kandy on their way home to Colombo. After all, our fabrication of being a married couple prevented us from telling them that Ann's actual destination was Colombo. And I had images of them remaining dockside to wave as we departed for the Arabian Sea and points beyond with Ann still aboard the *Verona*.

Figure 25. Fresco within the impenetrable rock mountain fortress to the left. Photos by author.

Determined protestations did not dissuade our kind hosts from delivering us to Trincomalee. But after fulfilling what they felt was their obligation, and following a brief tour of the *Verona*, they departed for home as Ann scrambled to transport herself and her belongings to the Colombo airport. We were thankful that our hosts had not exclaimed, "I have a friend in Trincomalee who owes me a favor" as we exited the subcontinent's version of *Groundhog Day*.

The evening of March 1, the *Verona* set sail for Malé, capital of the Maldives Islands, hundreds of miles to the west in the Indian Ocean. But my final night in Trincomalee was spent at a modest government-operated "rest home" in what formerly was a seventeenth-century Portuguese fortress set on a spectacular promontory overlooking the ocean. As the only guest in this very basic accommodation, I enjoyed the company of a kind elderly gentleman, Mr. de Silva, who appeared to be the sole employee on-site.

Mr. de Silva served as clerk, "bellman," waiter, and for all I knew, he was also the person who prepared my simple breakfast the next morning. I engaged Mr. de Silva in conversation because of his welcoming nature but also because I knew something of the fascinating history of Vasco da Gama's 1497 discovery of a sea route from Lisbon to the Far East. This maritime feat had propelled Portugal to a disproportionate share of European prosperity and power for a century or more, and my "host's" surname suggested that his family tree included Portuguese colonizers. I learned that he indeed was proud of his Portuguese ancestry, and that he and his family were Christians. He also spoke with pride of his children. I sensed that Mr. de Silva, frail and very advanced in years, was a lonely soul who craved human interaction. He also seemed very pleased when I asked if he would like me to take a photograph of him, promising to send him a copy. Because the photo was a slide, several months passed before I was able to make a print and mail it to him.

Days before Christmas 1966, the following letter arrived at my parents' home in San Antonio, adorned with two handsome stamps from Ceylon, one of which had the likeness of Queen Elizabeth:

Henry de Silva
Frotoft Group
Ramboda, Ceylon
15 December, 1966

"My dear Mr. Cline,

Received my daddy's snap with many many thanks. Before I start with anything else, I would like to inform you about a very sad news. My beloved daddy passed away on the 20th of November and on the same day I got his snap what you sent, from Trincomalee where he was working.

Now I would like to give you a self introduction about me. I am his eldest son. I am working on the above Tea

Estate as a Clerk. The estate where I am working is only 10 miles from Nuwara-Eliya, the famous place where the Touriest spend their holidays when they come to Ceylon. The place is very cold. I think you too would have visited this place when you were here.

As this is my first letter to you, I am not going to continue this letter any further. I will give you some more details in my next letter. I shall be very much thankful to you if you could kindly send me a snap of yours when replying. Also I would like to have another snap of my daddy's if I am not troubling you. The gift I can send you from Ceylon is only some good teas. I will send you some good tea very soon.

Wish you all a VERY HAPPY CHRISTMAS.

With Best Wishes.

Yours sincerely,

Henry de Silva"

Henry de Silva's letter touched me deeply. Now, after more than fifty years, I recall the emotion I experienced when reading it. I think about the final photograph of a gentle man and of its meaning for a grieving son. Of course I sent Henry the "snaps" he requested. I also let him know that while I appreciated his offer to send tea, it was not necessary.

My hope is that this short account of senior Mr. de Silva in Trincomalee will move at least one reader to explore the remarkable history of Vasco da Gama's voyage of exploration and its lasting global impact.

Nine

The Lost Treasure of Atahuallpa

Stellan Moerner was the sole crew member not born and reared in America. He came from a titled Swedish family, but more than citizenship separated him from the rest of the *Verona*'s crew. Older than most and lacking common ground and shared culture, Stellan was considered aloof and something of a mystery man and playboy. In his early forties, his lanky, erect frame, Nordic features, and deeply accented voice did nothing to diminish the gulf between him and his crewmates.

I didn't find Stellan such a mystery. In fact, we developed a firm friendship and shared a number of enjoyable onshore adventures before unfortunate circumstances led him to bid farewell to the *Verona* long before the sailing year ended. For Stellan, traveling the world was just another adventure, one of many that formed successive chapters in his life's story. Educated in Sweden as an engineer, Stellan accomplished just about everything a young European man of privilege would anticipate, with the exception of holding a bona fide job. A tradition in European aristocratic families, I am told, is to find creative ways to distance their "black sheep" from the ancestral home. In Stellan's family's case, the necessary distance was from Stockholm to Las Palmas on Gran Canaria, Canary Islands (Spain), a popular tourist destination off the coast of northwestern Africa. Stellan's cash cow (generating pesetas) in Las Palmas was a well-known watering hole, Bar Columbus. From this base of operation, and perhaps assisted by an income stream from family, Stellan was free to pursue his passions. One of them was beautiful young women. Successive marriages to Swedish actresses, some stars of cinema, ended in failure, apparently before children complicated the relationships.

But Stellan's most compelling and long-lived passion was his determined search in Ecuador for the lost treasure of Atahuallpa, a powerful Incan king. He spent perhaps a decade in this endeavor, which included leading arduous expeditions in treacherous mountains that routinely claimed the lives of the unprepared.

Francisco Pizarro, an illiterate conquistador from Extremadura, Spain, landed on the coast of Peru in late 1532, shortly after a massive civil-war battle in which Atahuallpa's forces defeated his brother Huáscar's army, leaving an estimated thirty-five thousand dead on the battlefield. Atahualpa, curious and apparently fearless, agreed to meet Pizarro and his fewer than two hundred men in Cajamarca (a city in today's Ecuador) on November 16, 1532. Massive chaos ensued, leaving Atahuallpa a prisoner of Pizarro, whose quest for gold and silver was legendary. In exchange for his freedom, Atahuallpa offered to fill a room of seventeen by twenty-two feet with gold, and two similar rooms with silver. He ordered his generals to carry out this task. Once accomplished, instead of his freedom, Atahuallpa was rewarded by being choked to death by his captors and then burned at the stake in Cajamarca's city square.

One of Atahuallpa's generals, Rumiñahui, according to legend, learned of his murder while on his way to Cajamarca with an additional load of treasures being transported by fifty thousand porters. Rumiñahui hid the treasures in a cave or in a lake in the Llanganati Mountains near Quito. Legend also tells us that Rumiñahui continued fighting the Spaniards and was captured and tortured but never revealed the location of Atahuallpa's treasure.

Stellan was one in a long line of dreamers who for centuries have searched for this treasure. Each had been convinced he had found the key to unlock untold riches and fame. Stellan planned to return to Ecuador after leaving the *Verona,* and repeatedly he encouraged me to join the expedition. The website www.thelifeo-fadventure.com/inca-gold/ not only tells the intriguing stories of treasure seekers over the centuries, but also offers a chain of communications that indicates that the search actively continues. Stellan was not the last to be seduced by Atahuallpa's lost treasure.

The first was apparently Valverde, a Spaniard who, according to legend, after marrying an Incan princess from the Llanganatis, suddenly became wealthy and returned to Spain. Later, near death and attempting to benefit his wife, he prepared a document known as Valverde's *derrotero* (path) describing the landmarks needed to find the remaining treasure, from which he had removed only a few pieces. After his death, Valverde's *derrotero* was bequeathed to King Charles V of Spain, who subsequently ordered an expedition. It seems that after something encouraging was found, the expedition's leader, Father Longo, disappeared under mysterious circumstances.

The next search took place in the late seventeen hundreds under the direction of Don Atanasio Guzman, a miner familiar with former Incan mines in the Llanganatis. He also disappeared, but not until he had prepared a detailed treasure map.

In 1860, English botanist Richard Spruce claimed to have discovered Valverde's *derrotero* and Guzman's map in archives in Latacunga, a town near the Llanganatis. Publication of his findings in the journal of the Royal Geographical Society precipitated an avalanche of British treasure seekers, including Barth Blake and George Edwin Chapman. A letter from Blake to friends stated:

> It is impossible for me to describe the wealth that now lays in that cave marked on my map, but I could not remove it alone, nor could thousands of men…There are thousands of gold and silver pieces of Inca and pre-Inca handicraft, the most beautiful goldsmith works you are not able to imagine, life-size human figures made out of beaten gold and silver, birds, animals, cornstalks, gold and silver flowers. Pots full of the most incredible jewelry. Golden vases full of emeralds.

Chapman didn't survive the journey out of the mountains, and Blake fell overboard on a trip to North America to sell the gold they'd taken from the cave. Not surprisingly, the search for the treasure has also been closely associated with "the curse of Atahuallpa's gold."

You've already read of some of the victims of the treasure's curse: Father Longo, Guzman, Chapman, and Blake. But that's just the tip of the iceberg.

In the mid-1930s, a Scotsman named Erskine Loch mounted two disastrous treasure hunts in the Llanganatis. During the first expedition, porters deserted Loch, and violent rains dogged him for thirty-seven out of thirty-nine days. On his second trip, Loch's party ran out of food and fell to hallucinations. "The country ahead," Loch wrote in his book, titled *Fever, Famine, and Gold*, "had spur after spur of precipitous rock faces descending into raging torrents below. Everything we stood upon, everything we clutched gave way under us." Soon after the book's publication, Loch shot himself.

Yet others kept coming—and dying. In the 1920s, an American known in local accounts as "Colonel Brooks" established a bank in Ecuador and then got the treasure bug. On his first trip into the mountains, his porters mutinied. Later, Brooks decided to take his wife, Isabella, to the Llanganatis for a "romantic getaway," but they were promptly greeted by torrential rains. She died of pneumonia, and he ended up in a madhouse in New York—muttering wildly, one imagines, about gold and silver and emeralds.

Bob Holt, an American geologist from Arizona who had worked with various oil and gold-mining companies in Ecuador, on his first expedition into the Llanganatis slipped and fell on a sharp broken tree trunk. It stabbed him directly through the heart.

If you feel compelled to read a serious, detailed, and objective account of the efforts to locate Atahuallpa's treasure, I recommend *Valverde's Gold: In Search of the Last Great Inca Treasure,* by Mark Honigsbaum. I was excited when it arrived via Amazon at my post office box. And I almost jumped out of my skin after immediately turning to the index and finding "Moerner, Stellan, 134, 311, 312." With the exception of a letter I had received from Stellan from Las Palmas in late 1966, this was my first shred of information about him since he decided to leave the *Verona* in Hong Kong.

On page 134 of Honigsbaum's book, I read:

> After the publication of the first edition of his book Andrade continued to delve. He discovered that in 1812 General Toribio Montes, president of the Audiencia of Quito, had also organized an expedition to the Llanganatis. However, the papers relating to the expedition had been bought by a Swedish count named Moerner—one of the growing band of Llaganatis enthusiasts—and he wouldn't agree to share them.[1]

Then on pages 311–12, I found this:

> In the late 1960s, armed with the general's papers, Moerner had secretly journeyed to the Llanganatis to look for the

1. Mark Honigsbaum, *Valverde's Gold: In Search of the Last Great Inca Treasure* (New York: Picador, 2005), 134.

> Incan gold mine. Whether he succeeded in finding it no one could be sure, but in 1970 he'd been photographed on the Río Topo holding what appeared to be two massive rocks of gold ore. On his return to Europe, Moerner sent the photograph to *The Times* of London, telling the newspaper he had found the Inca mine and was planning to go back to Ecuador shortly to recover the treasure. To finance the expedition, Moerner established an investment vehicle, The Llanganati Treasure Society, and issued two thousand shares valued at 500 Swedish kronor each. Subscribers, promised Moerner, stood to make back "ten times" the value of their investment.
>
> *The Times* obligingly published Moerner's fantastic story, together with a picture of the share certificates, and within the year he was back in Ecuador. But Moerner never launched another expedition, taking a suite at the Hotel Quito instead, where he drank away his investors' money.
>
> "What became of him?" I asked.
>
> "The last thing I heard was he was running a hotel in Gran Canaria."[2]

I pondered this not-completely-surprising revelation about my former shipmate. I wondered if perhaps Stellan was a gifted con man from the beginning and whether his noble title was part of his confidence game. During a brief stay in Stockholm during the 1980s, I called the two Moerners listed in the city phonebook, without success, but that was the extent of my effort. But then I recalled an event in Papua New Guinea that seemed to confirm that Stellan was a well-known "celebrity" in his native Sweden. The *Verona,* having arrived off Madang, Papua New Guinea, glided into its assigned mooring near the gleaming modern Swedish cargo

2. Ibid., 311–12.

Figure 26. Stellan Moerner, highlands of Papua New Guinea. Photo by author.

ship the *Tenos*. Stellan and I were on deck as we arrived…and his eyes widened when he spotted the ship flying the flag of his native country. He said to me, "Barney, let's go have some fun!" Once cleared to disembark, in the early evening, the two of us walked to the cargo ship, and Stellan started calling out, in sing-song Swedish, of course. Soon someone responded. Stellan introduced himself, and the ship's captain was informed. After a brief pause came an invitation to board. Stellan was greeted enthusiastically, and soon we were feasting with the officers and the *Tenos*'s captain: Swedish breads, butter, herring, cheese, reindeer meat, two kinds of aquavit, white wine, burgundy, coffee, and brandy. Leaving the *Tenos* about midnight, carrying gifts of cans of herring, aquavit, and cigars, we were invited to return in the morning to join the captain for breakfast. Without question, Stellan had received a royal reception, was treated with extreme respect, and was offered an extravagant array of Swedish delicacies from the captain's special chest. While I understood virtually none of the spirited conversation during the leisurely breakfast of fish and herring and poached eggs and potent alcoholic beverages, it was obvious that the captain was deeply interested in his honored guest. This experience convinced me that Stellan was indeed a Swedish count, widely known in his country.

Did I see any clues that would portend the fraudulent activities described in Honingsbaum's book? During long stretches at sea, the two of us often played gin rummy for very modest stakes. Repeatedly, I detected Stellan's inclination to cheat...which I accepted as forgivable moral lapses. Was this behavior handwriting on the wall? Perhaps.

The *Verona*'s youngest crew members did not take well to Stellan. As indicated earlier, he was considered aloof. He could also be demanding and had minimal compunction about shirking duties. While on watch, he would occasionally perform expected chores, swabbing the deck, scraping paint, painting, cleaning brass. But often he would simply ignore the orders of his watch captain, driving the teenage males crazy. But his most egregious sins occurred with regularity at meals, served family-style, when he would attempt to spear the choicest piece of meat on the platter. This was unforgivable, unacceptable behavior at sea, where equitable sharing at the mess table is a paramount virtue.

So the teenage sailors crafted their revenge, giving Stellan a hard time at every turn, doing their best to make life miserable for him. And I suppose they succeeded, because one day he confessed that it was no longer fun for him and that he was going to leave at the next major port of call, Hong Kong. And he did.

The most potent weapon the young guys used against Stellan was to constantly refer to him as the "no-count count!" Perhaps they had it right all along.

Ten

A Lord in the Highlands of New Guinea

As one accumulates birthdays and decades of harmonious matrimony, it is easy to forget the inclination of single youths to share friends. This sharing is especially common among expatriates surrounded by a culture distinct from their own. I experienced a chain of friend-sharing events shortly after arriving November 17 in Madang, situated on the northern coast of Papua New Guinea (known since 1975 as the Independent State of Papua New Guinea). In contrast to most ports of call, Chris decided that the *Verona* would remain in Madang for a week to permit crew members so inclined to explore this unique land. This was welcome news for several of us.

A curious gentleman I met on shore, when he learned that I was the *Verona*'s doctor, called Bronlyn Vickers, wife of Australian physician Wesley Vickers. During that era, Papua New Guinea was administered by Australia, and consequently most medical professionals, including Wesley Vickers, were Australian citizens. Soon Bronlyn picked me up and took me to the hospital where Wesley was employed. The spontaneous hospitality of this fine couple overflowed such that by the end of our initial meeting, they had almost adopted me.

That next day, at the invitation of the Vickerses, Stellan Moerner and I took a taxi to the Vickerses' home, where we met Bronlyn's friend Linda Dadak from the UK. Linda had offered to stay with the Vickerses' children that evening while they were out, so Stellan and I volunteered to keep her company. Linda was keen to share suggestions about how to best spend our week of freedom, and we were eager to hear her ideas. She strongly urged us to fly into the central highlands, go to the town of Banz, and there to contact her English friend Miles Barne, who operated a nearby coffee plantation. She offered to send Miles a telegram to expect us and to ask him to meet us in Banz.

The following day brought a whirlwind of activity as we prepared to visit the fabled central highlands of Papua New Guinea, populated by "Stone Age inhabitants" unknown to the world not many decades earlier. During the evening, Stellan and I enjoyed excellent drinks, food, and wine with Wes, Bronlyn, and Linda at Smuggler's Motel in Madang. We returned to the Verona exhausted, armed with Linda's letter of introduction to Miles.

Leaving Madang on an Ansett-MAL DC-3 cargo plane, Stellan and I flew to the highlands and landed near Mount Hagen. A taxi took us into the town of Mount Hagen. We then found a ride to Banz, where we had our first up-close look at the native population. I was struck by their colorful dress, physique, and proud stance. Before Miles showed up, we somehow managed to meet Ian Rutledge, Holt Binney, Tony and Jan Bierne, and others with whom we shared sandwiches and beer. It became obvious that the far-flung expat network in the region was strong.

The drive through the Wahgi Valley with Miles took us through appealing mountainous countryside. Soon we reached his coffee plantation, Warawagi, with the rapidly flowing Wahgi River transecting it. Sadly, I learned, Miles's cousin Peter Maxtone Graham had drowned in this river the previous year. Miles, who had come out from England to Papua New Guinea and Australia three years earlier, went to Warawagi when he learned of the death of his cousin. His initial intent was to settle Peter's affairs and to leave, but the allure of the central highlands captured his imagination and led him to stay on. Peter had married a native woman, Gol, and they had two children, Peter and Jamie. Later, with his second wife (Nun Waim), daughter Barbara was born. Barbara was a toddler at the time of my arrival. Concern for the welfare of Peter's children contributed to Miles's decision to remain at Warawagi.

Tall, lanky, a few years younger than I, and with a confident, reserved nature, Miles encouraged me to join him for a swim in the Wahgi River shortly after our arrival. This turned out to be a brisk, invigorating one-mile float. At some five-thousand-feet elevation and delightfully cool, the vistas from this remote site were intoxicating. Miles lived in his late cousin's simple but cozy home, constructed entirely of local materials, with a high thatched roof and a sturdy rock fireplace in one corner.

A similarly constructed but smaller guesthouse offered up a wonderful night's sleep for Stellan and me. The morning was cool and refreshing. Stellan and I relaxed and read in the sun until relishing a late breakfast. Having already been introduced to the plantation's solitary outhouse the night before, it was not until morning that I fully appreciated it. I was reminded of the familiar adage "location, location, location" when, seated upon the throne with the door open, I was treated to a magnificent panorama with the Wahgi River in the foreground and sprawling mountains behind.

That afternoon, Miles drove us to the Hallstrom Trust Park in Nondugl, a bird-of-paradise sanctuary. While few birds were to be seen, we did meet Robin and Pat, who were staying with Bill and Ann Stokes. They (the Stokeses) were Miles's closest friends in the area, and Bill, an ex-patrol officer, was also his partner in a camera shop. As we returned to Warawagi via the southern side of the Wahgi Valley, Miles described the three main subgroups of Europeans in that region of New Guinea: missionaries, government

Figure 27. Miles Barne surveying his Warawagi coffee plantation in Papua New Guinea. Photo by author.

employees, and planters (or other entrepreneurs engaged in free enterprise). We also discussed the built-in antagonisms inherent in this tripartite mix of expatriates. Some of these antagonisms I had recognized during the *Verona*'s stops in the New Hebrides (now Vanuatu). Miles described to me the three principal sources of conflict, and sometimes violence, within the native population: land, pigs, and *meri* (local pidgin for woman or wife).

The evening was spent by the fireplace learning more about the fleeting "golden period" being experienced at this stage in the evolution of Papua New Guinea's highlands. Except for the few clusters of houses around church-sponsored missions, there were virtually no villages. Rather, houses were widely scattered. Typically, they were well constructed and had good gardens, with sweet potatoes the main crop. The recently arrived Europeans remained a novelty to the indigenous inhabitants, whose social and cultural fabric remained intact. Consequently, the relationship between the two groups was rather harmonious, characterized perhaps as one of mutual fascination rather than overt domination or submission. Not surprisingly, this delicate balance has not persisted as national pride, exploitation, and other sources of antagonism have increased in the region.

After a quiet evening by Miles's fireplace and another good night's sleep, Stellan and I accompanied Miles to Banz while he shopped for supplies the next day. In the afternoon, we took a long drive in the Danga country, where, on a remote stretch of road, we approached a handsome local tribesman wearing his eye-catching dress, smoking a cigarette. Hanging from his nose was a circular shell disk and from his belt hung a massive array of shells, a cascade of ankle-length strips of fur, and a metal axe. Additional adornments included a broad neckpiece carved from a large shell, woven arm bands above his elbows and wrists, and a bird-of-paradise plume that flowed majestically from his hat. This proud man had no objection to me taking his photograph.

Figure 28. Danga tribesman, Papua New Guinea. Photo by author.

That stimulating evening (the November 22 anniversary of the death of John F. Kennedy) was spent at the home of Bill and Anne Stokes. A fine supper was followed by spirited conversation, slides, and finally a movie. We didn't arrive back at Miles's plantation until after 0200. Happily, twenty-four years later, living in New Orleans, Nancy and I were able to reciprocate the Stokeses' warm hospitality when Anne, by then widowed, visited us for several days.

The next day was memorable for two reasons. First, a traditional "sing-sing" to welcome us was held by the Konambuga tribe. This spell-binding event consisted of dances accompanied by chanting the story of our arrival and the presentation of gifts: cigarettes, lemonade cans, potatoes, onions, chickens, eggs, and more. But the most appreciated gift was the photographic opportunity...as I raced through several rolls of Kodachrome film. Each "dance" had a theme. For example, one focused on fire starting and featured a tribal medicine man I called the Jolly Green Giant (photo below). Painted from head to toe with greenish-white clay, he wore a fanciful headdress, and his torso was obscured by plant leaves. His private parts were noticeably contained within a tin can, considered an important status symbol. Then a vigorous "war dance" was performed by warriors with imposing physiques, painted faces, and elaborate dress, who thrust spears, and brandished swords as they shouted taunts at an imaginary enemy. The head of each warrior was topped with a headdress exuberantly displaying highly valued bird-of-paradise feathers, and almost every warrior's nose was painted white. A subdued "courting ceremony" called *karim leg* consisted of two handsome young couples seated with their legs crossed. It was explained to us that this practice was akin to holding hands while watching TV in America.

Second, as though the day had not been sufficiently unique, Papua New Guinea in the late afternoon was treated to an 80 percent eclipse of the sun. I half-expected the indigenous people to

Figure 29. Left: Konambuga warriors in full regalia; right: the "Jolly Green Giant". Photos by author.

be unsettled by the strange light it cast, but they appeared either unaware or unconcerned. That was our final night at Warawagi.

The next day, Miles drove us to Goroka, from which our flight to Madang would leave the following morning. The drive took us through the Wahgi and Chimbu valleys. In the latter, the local population was much less colorful than in the Wahgi. In Kundiawa, we stopped for refreshments at the Lutheran mission, and then Miles deposited us at a Goroka hotel before returning home.

Back in Madang, there was much to relate to Linda and to Wes and Bronlyn about our central-highland adventures, and tea at Linda's home and leisurely lunch and dinner at the Vickerses' provided this opportunity. That final evening was topped off with a viewing of the movie *Goldfinger* and then coffee. Farewells did not come easily at noon of November 27 as the *Verona* departed on what would become a sixteen-day passage to our next port of call, Kaohsiung, Taiwan. Nor did we know that the day before reaching our destination we would battle gale-force winds in the South China Sea.

New Guinea reappeared on my mental screen nine months later, in New York City. After the *Verona,* I had returned to my parents' home in San Antonio for a short visit. My interest in tropical diseases and public health had been steadily growing since a medical-school fellowship in Panama, two years with the Peace Corps, and my year on the *Verona.* I therefore applied to and was accepted into a Master of Public Health degree program at Johns Hopkins University in Baltimore, to begin in September 1966. And with lingering misgivings I notified Tulane Medical School that I would not enter the surgical residency. I had time on my hands before moving to Baltimore and decided to earn some money by spending six weeks covering the busy private medical practice of a general practitioner in San Marcos, Texas. He was finally taking his first vacation after twenty-five years of practice.

A few weeks after moving to Baltimore, I went to New York to visit an old buddy, Ernie. We had been friends since college days. Ernie, a fine-looking hunk and former basketball star standing six feet six, had no shortage of girls interested in him. One, Norma, had enjoyed a long, close relationship with Ernie, the sort that seemed destined to end in matrimony. I had known Norma for years, considered her terrific, and had hoped that Ernie would pop the question. But that had not happened, and as Norma waited, her frustration grew, and she shared with me that she had decided to travel around the world for three months to "get away from Ernie for a while." She worked for a travel company that allowed her to take the necessary leave.

Norma was (and remains) a high-spirited, adventuresome soul and very dear friend. When I learned that her travels would include Australia, I encouraged her to visit Miles and to see Warawagi and the central highlands of New Guinea. She agreed to this "side trip," unaware that Miles was an eligible young bachelor. I may have left her with the general impression that he was an English planter, and I suspect she visualized him as a portly, fiftyish, mustached gentleman wearing a pith helmet. I wrote Miles and told him to expect a visit from Norma. As an astute reader, you have likely anticipated the outcome. The first letter I received from Norma from New Guinea conveyed rather positive impressions. For example, "Well, here I am at Warawagi and I must say it is even more beautiful than I thought possible." Also, "Miles is really quite special and don't breathe a word of this but if it were not for Ernie at home I would stop my trip at this point and stay here much longer." Two weeks later, her letter read, "Well, what can I say? I'm still at Warawagi and will be here for at least another three weeks I imagine" and "Love this place and the people and really can't think of leaving. Everything is quite confusing in my mind and I wish I knew what was happening with Ernie." And months later:

> Guess what? Miles and I are engaged to be married. I don't believe it myself. Miles was driving me to the airstrip in Banz to get my ticket and just before we got there he proposed. We were laughing, hugging and kissing each other as we drove all around the strip. We decided not to say anything until the night before I was supposedly leaving. At 7 PM Miles made the most priceless speech, toasting Barbara's birthday, Warawagi hospitality, England and her colonies, the Queen, America, and then in closing he said that for the cause of his comfort and Anglo-Colonial friendship he bid all to drink to his future wife Norma.

Of course, it was painful for me not to be able to attend their wedding, which took place on the airstrip in Banz.

After leaving Baltimore, a year later, living as a doctoral student at the University of California, Berkeley, a letter from Norma informed me that she would be coming through San Francisco on the way to visit her family in New York and that with her would be Barbara. We enjoyed a short but sweet visit, and I could see that Norma was radiantly happy with her new life far from New York. Nancy and I were married not long after Norma's visit, and we kept in touch with her as the years crept by. She and Miles visited us in New Orleans in the late 1970s, but it was not until June 1984 that we again were together. With our then ten-year-old son, Philip, and seven-year-old daughter, Lea, we made our first trip to Australia. After some days of enjoying the cosmopolitan charms of Sydney and the outgoing character of Australians, we flew to the small town of Moree in New South Wales. Miles and Norma by that time had traded New Guinea and Sydney for the Australian outback, where they lived with their two sons, George and Thomas, just a bit older than our children. Miles had acquired a large cotton-farming operation near the village of Biniguy. The farm was home to an array of massive farming equipment, a handsome sprawling farmhouse with endless porches, a four-seat Cessna, and a beautiful white pony named Whiskers. They even had their own abattoir to ensure a supply of fine beef in this rather remote area. Norma, ever enterprising and fearless, had completed a six-month butcher's training program in Brisbane so that she could expertly supervise the abattoir's operation. From their farm, a visit to a neighbor or to a community social activity, such as a horse race, involved a flight in the Cessna with Miles at the controls. Roads were few, and distances great.

Their lifestyle had certainly taken a step up from Warawagi; for the first time, we understood that Miles was a man of means. At some point, Norma had told Nancy and me that Miles, as the eldest son in his British family, was expected some day to return to manage the family holdings and the manor. But it was many years later, after their marriage had ended, and she had moved back to Sydney, that we more thoroughly digested this information. Norma, visiting us in Texas, had briefly described Sotterley in England. During periodic visits with Miles to this Barne family home, Norma had helped entertain the locals while Miles took care of other demands. By the time of Norma's Texas visit, we learned that after the death of his father, Miles had moved to

Sotterley and had remarried. His new wife, Tessa, had two daughters from a previous marriage.

Periodically, Norma and, separately, Miles encouraged us to visit Sotterley, a name meaning "southern meadow," derived from the Saxon language. Not until 2012 did our first visit take place. While we had not seen Miles for some twenty years, we remembered his intelligence, hospitable nature, and keen fascination with history, but we had no clear idea what to expect when Nancy and I boarded a train in London for the 120-mile trip northeast toward Suffolk and the North Sea. We changed trains in Ipswich, and shortly afterward, Miles met us on arrival at the tiny Brampton railway station and welcomed us warmly. Wearing an aged wool sweater and working attire, he lifted our bags into the back of his beat-up old Range Rover as we prepared to drive the ten miles or so to Sotterley. Suspense built as we neared a rear entrance and parked in a massive ivy-covered barn-like brick structure and were greeted by Miles's two adoring dogs. As we passed through a door and into light drizzle, our jaws must have dropped as we realized that before us lay a manicured country estate and an immense Georgian manor house of the sort that one sees in movies or in news-clips about the British royal family.

Figure 30. Sotterley Manor House, Suffolk, England. Photo by author.

Entering the house, we were greeted by Tessa, who won our hearts then and has continued to do so with her gracious nature, wit, unpretentious ways, and upbeat view of the world. Gradually, we were able to piece together our evolving perception of Miles, having progressed from struggling coffee farmer in Papua New Guinea to prosperous cotton farmer in New South Wales to lord of the manor of Sotterley Estate. Over time, including the Barnes' visit with us in Texas in 2014 and our repeat visit to Sotterley in 2015, we came to better understand and appreciate his fascinating world, the Barne family history, and the influences upon him of powerful British traditions.

As the eldest son, Miles was obliged to oversee the operation of the estate after the death of his father, Michael Barne. After all, the estate had been in his family since 1744, when a much earlier Miles Barne purchased it from John Playters and undertook to rebuild and expand the existing manor house. (That early Miles Barne was descended from Sir George Barne, lord mayor of London in 1552.) I also learned about the earlier history of Sotterley (which has been spelled in over fifty ways). In 1086, during a national census of landowners in Britain, the village was recorded to have twenty-two inhabitants, but over the centuries, it grew to over three hundred. Currently, the count is about one hundred. The first lord of the manor was probably Buchard, a Saxon, before the Norman domination of this region. The present (Norman) Christian church was built around 1300 by Roger de Sotere on the site of an earlier Saxon church, so it has been a continuous place of worship for the community for well over nine hundred years. The record of subsequent lords of the manor is not easy to decipher, but apparently in 1479, the Satterlees sold the land to the Playters family. Sir Thomas Playters (a brass portrait of whom is found in the church) assumed the role of lord of the manor, and the estate remained in the Playters family until Miles Barne, a member of Parliament, purchased Sotterley in 1744 (decades *before* American colonists declared their independence from England). I make a point of including this history because it is not easy for citizens of our "nation of immigrants" to appreciate the family and patriotic obligations Miles must factor into his life choices.

Nancy and I were educated and entertained with stories both remote and recent as we walked with Miles through the house and gardens and fields of the five-thousand-acre estate. We appreciated the beauty and the time lines preserved in and around the church and on the grave markers. I felt as though I was enrolled in a crash course in medieval art, architecture, and history. In the church, I queried Miles about a stained-glass window inscribed "Restored by Mabel Satterlee Ingalls of Sotterley Maryland, USA, 1953" and learned that members of the Satterlee family had emigrated to America in the late sixteen hundreds and later had founded Satterlee Plantation in Maryland. I learned that a contingent of Satterlee family members had visited Sotterley in 1891 and that a few years ago a similar group of Satterlees was welcomed to Sotterley Estate by Miles and Tessa. The following year, Miles and Tessa were guests at the Maryland plantation.

Also in the church, I read a metal plaque for Miles's father with the following inscription: "In loving memory of Michael E. StJ. Barne, Lieutenant Colonel Scots Guard, eldest son of Major Miles Barne DSO. He lived at Sotterley for 50 years caring for his estate and working in his woods which he loved. July 1905–March 1979." We also learned that during World War II, because of Sotterley's proximity to the North Sea and the threat of invasion by Germany, the estate had been taken over by the British military. With his father away in the war, Miles's mother took charge. Before the British Army moved in, she sealed off the wine cellar with a sturdy brick wall, and she managed to move valuable items and some furniture into the locked attic on the third floor. Italian prisoners of war were among the first to occupy Sotterley, under guard. Miles remembered that they deeply cared for their "prison" and strove diligently to maintain the gardens and orchards and to treat the house and its contents with the utmost respect. Some returned to Sotterley years later to proudly show their wives and children where they had spent most of the war. Miles related that in stark contrast, the British Army troops typically trashed the place, often treating it with scorn. Miles also pointed out to us a giant tree in the middle of a big pasture, which, during World War II, had been used as a navigation point for bombers returning to their home bases after missions over the continent.

The manor house is a living museum. The walls offer a succession of portraits of ancestors, some sternly overlooking the cavernous sitting rooms in which they hang. When asked, Miles patiently introduced some of them to me and Nancy. At one portrait, I recall, Miles commented that this particular ancestor in the late seventeen hundreds was a partner in establishing Lloyd's of London, perhaps the world's best-known and trusted insurance company. This fact seemed to dovetail nicely with other information I have gleaned about the Barne family from a history book. In an era in which England was emerging as a major European power, the family was deeply involved in the trading companies instrumental in the colonization of what became the United States of America. The Barne family was

> invested, in varying degrees, in the Muscovy Company (1555), the Eastland Company for Baltic Trade (1579), the Levant Company (1581), forerunner of the East India Company (1600), the Virginia Company (1606/07) and the Hudson Bay Company (1670). Longer term many of them formed the basis of the British Empire.[1]

---

1 Richard Lloyd, *Welcome to Sotterley and Its Important Transatlantic Connections* (Hough on the Hill, Lincolnshire: Barny Books, nd), 9.

So it is reasonable to surmise that this family played a significant role in our colonial past.

Near our bedroom on the second floor was a display case with a stuffed penguin, still handsome but in need of restoration. Asking Miles about the penguin, we learned about his great-uncle Captain Michael Barne, who was with the famous explorer Scott on HMS *Discovery* during Scott's first expedition to explore Antarctica. More information about Captain Michael Barne and his exploits are chronicled in the excellent book *Shackleton: By Endurance We Conquer,* by Michael Smith. Nancy roamed the attic of the manor house with Miles. Afterward, she described a mind-boggling depository of centuries-old furniture, art, and other items, such as guns and military apparel. Miles commented to her something to the effect that one day all this stuff needed to be sorted out...

Further, we came to appreciate that while Miles's position commanded utmost respect in the community, and while his assets were many, his was a modest, unpretentious, and hardworking lifestyle largely devoid of luxury. Simply the upkeep of the forty-plus-room manor house was an ongoing burden. For example, the year prior to our first visit, they had replaced all the floor-to-ceiling massive window frames and windows, an enormous and costly undertaking. The next year was spent replacing the entire roof of oak shingles. Lacking central heating, the only room in the house that remained warm and cozy in the cold months was the kitchen, where they spent much of their time. Miles, with his long frame, offered that the only "luxury" he allowed himself was to travel business class when he flew to Australia.

While Miles is responsible for overseeing all of the estate's business affairs, his own special area of interest and expertise is forestry, a field in which his father had also excelled. He is an active member of the Royal Forestry Society dedicated to forestry. With him we admired giant oaks, some of which were known to be over five hundred years old. With pride and knowledge based upon the best available science, he showed us fields of recently planted oak saplings that would be thinned in six to seven years to retain the prime young trees that would not be harvested for timber for at least one hundred years. He then led us to an area where mature trees planted by his great-grandfather had just been felled and cut into massive logs to be converted into high-quality timber. We also learned about the unfortunate introduction of North American gray squirrels into Britain in 1876, their rapid spread and displacement of the native red-squirrel population, and their catastrophic impact on the timber industry. It seems that gray squirrels, more so than their red cousins, like to strip tree bark, causing great damage. In the woods, Miles introduced us to his full-time employee Fred Mallett, known as "Jigger," his only job being to hunt and trap gray squirrels. Miles's comment about Jigger: "He has a heart of gold, is an observant naturalist as well as a ruthless destroyer of vermin."

The estate has its own busy timber mill and drying sheds. On a typical day, Miles can be found hard at work in his estate office, attending to his forestry demands or responding to a range of other challenges encountered in a rural environment. We also learned of additional sources of income for the estate, including properties that were rented for homes or stores. Miles employs a full-time property manager for this task. Also employed is an agricultural manager, who is responsible for overseeing a variety of crops, including rapeseed, wheat, carrots, and peas. Lush green pastures are leased to people who bring their sheep to graze, shear, and lamb. Licenses are sold by the estate for use of the estate's shaded horse-riding trails. The estate is even leased on occasion for traditional hare-hunting events, akin to fox hunts but using a breed of dogs known as harriers to pursue hares.

Following their 2014 visit with us in Texas, a return to Sotterley in 2015 was a treat for Nancy and me because of the opportunity to spend quality time with Miles and Tessa and to more fully understand and appreciate their lives in Sotterley. Our newly painted and redecorated bedroom on the second floor, spacious and stately and with gorgeous window dressing, offered a spectacular view of the grounds and gardens. Tessa had for some years been refurbishing bedrooms because family visitors were frequent, including their children and grandchildren. We had the pleasure of meeting some of Miles's delightful cousins, who spent the night at Sotterley after a nearby wedding celebration. During the evening meal of the final day of our visit, Tessa excitedly exclaimed that we must be certain to see *Downton Abbey* on TV that evening. It was to be the final installment of the season—not to be missed! Later, we all gathered in a cozy TV-viewing room with a fire blazing in the fireplace. Nancy had seen this popular TV program a few times and was familiar with it, but my awareness of it was minimal. Tessa was clearly an avid fan, and my impression was that Miles's attitude was one of detached indifference. I politely sat and watched as a rather strange sensation slowly crept over me. Here we were in surroundings identical to Downton Abbey watching *Downton Abbey* on TV!

Figure 31. Miles and Tessa Barne in kitchen of Sotterley Manor House. Photo by author.

In August of 2016, Miles's son Thomas, with his wife and two children, moved from Australia to Sotterley to take over from his father the daunting task expected of Barne male offspring. Miles and Tessa have turned over the manor house to them and have moved into adjacent quarters. I suspect that before many years pass, Thomas will be busy planting trees to be harvested in some 150 years by Barne descendants.

Eleven

The Sexual Life of Savages

I confess to having qualms about describing Vic Busuttin, the most unique person I encountered during my year at sea. In today's world, in our culture, he would be considered a pedophile, a sexual predator, a molester of young girls (albeit with the explicit consent of their parents). Our justice system would certainly deal harshly with Vic, securing him behind prison walls. Yet in the Trobriand Islands, Vic was a treasure, warmly welcomed by the very families of the girls that were offered the opportunity to spend a year or more at his home.

Our seven-day passage from Espiritu Santo (then the New Hebrides, now Vanuatu) began the day following Halloween 1965. The morning of November 8, the *Verona* anchored off Samarai Island. Located near the southeast extreme of New Guinea (the world's second-largest island after Greenland), Samarai Island was discovered in the late nineteenth century and for some decades became an important trading post in the region. At the beginning of World War II, it was largely destroyed to prevent its facilities from falling into the hands of advancing Japanese forces. At the time of the *Verona*'s arrival, the town, while an administrative and commercial center, certainly would not be described as bustling.

I was enthusiastic about reaching this remote land. Not many years earlier, mountain tribes, living under Stone Age conditions, were being encountered for the first time by European explorers and anthropologists. After my customary practice of collecting and sending mail from the post office and visiting the hospital to meet the local physician, not far down my "to do" list were creature comforts such as ice cream or cold beer. The evening of our Samarai arrival, I joined our cook, John Narayan, to visit a freighter with an Indian crew, where an abundance of spicy food and refreshing drinks was consumed. In his exuberance that evening, John praised the Taj Mahal, saying that when we reached Madras (now Chennai) a few months hence, we must visit it together.

Around noon the following day, I wandered into the Samarai Club, a pub of sorts near the waterfront where expats congregated. As my eyes adapted to the relative darkness, Vic appeared at the end of a wooden bar, a solitary figure sipping his beer. I estimated his age to be around seventy and noted his long wiry frame, massive hands, and prominent beak of a nose. Also revealed was his outgoing nature, for without hesitation he beckoned me to join him. Of course, locals in a port town like this are acutely aware of newly arrived vessels, especially one as distinctive as the *Verona*. Vic had quickly pegged me as one of its crew members.

Over a simple lunch and beer we chatted. I learned that Vic, originally from Australia, had spent most of his life in this part of New Guinea. Among other occupations, he had been the captain of a coastal freighter, and during World War II, he had served in the Australian navy. Initially, I was skeptical about this odd, witty, tale-spinning character with an Australian accent so thick that at times he was unintelligible, but the more I listened to him the more I was convinced of his unique knowledge of this remote land. When I mentioned to Vic that our next destination was the Trobriand Islands, his eyes lit up as he described how he knew these islands like the back of his hand. "My very favorite islands," he gushed. My immediate thought was to introduce Vic to Chris because he might offer valuable advice.

Like a schoolboy on the verge of vacation, Vic jumped at my suggestion to meet our skipper. Soon, Chris and Vic were chatting and to my surprise, Chris invited him to join us as we sailed to the Trobriands. Agreeing immediately, Vic explained that the *Verona* would need to stop for an hour or so at his home to pick up a few items. Vic also stressed the need to buy a supply of "stick tobacco" before departing Samarai. Tobacco sticks, he explained, were plentiful and inexpensive and served as the universal currency in this part of the world. He led me to a nearby trading store that had stick tobacco, crudely wrapped in packs of ten, and other unfamiliar items.

Leaving Samarai about 17:00 on November 9, we stopped for Vic to gather clothing and personal items at his home. By 20:30, we were again underway. Oceanic sailors on long voyages are typically pleased with the diversion offered by newcomers. The *Verona*'s crew welcomed this ancient mariner with minimal hesitation and seemed eager to absorb his colorful tales. Vic ended up sleeping in the empty bunk above mine in the forecastle, so consequently he became my garrulous companion for the next several days. This arrangement offered me ample opportunity to learn more about him and about this part of Papua New Guinea. Some of the things I learned startled my rather young, relatively unworldly ears.

Originally from Australia, Vic claimed that his family owned Brampton Island off the coast of Queensland. In *Wikipedia,* I found the following when searching online for Brampton Island: "In 1916, Joseph Busuttin and his family became the island's first European settlers. The island's resort was first established in December 1933 when two of the Busuttin's sons welcomed passengers from the P&O ship *SS Canberra.* Busuttin's sons then sold the resort and left in 1959." I surmise that Vic was one of those sons.

Remarkably fit, powerful, and agile at age seventy-five, Vic had spent most of his life as a trader, seaman, freighter captain, hunter, adventurer, and collector of young girls. His mother was still alive at age 105. A compulsive storyteller, Vic related to me that at his home he "hosted" about ten to twelve young girls, most of whom stayed with him for a year or two before returning home to their Trobriand Island families and that he would be visiting some of their families on this trip. Many of his "guests" were apparently prepubertal because, as he explained, he would use this visit to scan for new such candidates. And it appeared that his sexual prowess was that of a young man, seeking daily gratification. Initially, I maintained a degree of skepticism about Vic's stories, but there were no facts I could refute.

Opening his rudimentary travel bag to show me the contents, Vic explained again that sticks of tobacco serve as the common currency on the islands because coins and paper money have no meaning to these isolated populations. His one hundred or so tobacco sticks, tied into small packets, may have represented a small fortune for Trobriand Islanders. Beads and other trinkets, such as small mirrors, he related, were also handy items for trading.

Under sail for the Trobriand Islands, the morning of November 10, we anchored near a plantation at Esa'ala owned by Vic's friend, Australian Norman Evennett. Norman transported Vic and me to shore and introduced me to his wife, Mona, son Jeff, and house girl Marie and her baby. With pride, Norman showed us his remarkable collection of seashells before Chris and other crew members arrived to enjoy sandwiches and snacks. Afterward, we all walked to a nearby Methodist mission and "bush hospital."

A throng of local inhabitants at the mission asked to visit the *Verona,* but as their visit began, Norman grabbed my arm and suggested we quickly escape to the quiet and comfort of his home, where Vic had remained. I had so enjoyed my brief, refreshing time at their home that I took from my cabin a few gifts, treats

I had acquired in Samarai for my enjoyment: a bottle of good sherry, a pound of tasty cheese, and crackers. This gesture must have been greatly appreciated because later Vic shared with me Norman's comment that I was the first visitor ever to do this. The afternoon spent with the Evennetts was memorable. Despite their isolated location, I learned that Norman and Mona were exceedingly well-read and informed of current global events, pursued a broad range of interests, and had two sons at a school in Australia. A splendid meal and sumptuous pineapple pie with ice cream were accompanied by an enchanting rising moon. Reluctant to see this fine day come to a close, I ended up sleeping on the *Verona*'s deck because the air was very still that night.

The next morning was November 11, Armistice Day (now Veterans Day), 1965, my mother's birthday. I had sent her a telegram before leaving Samarai. Soon I returned to Norman's with Vic, Lloyd, Edgar, and Sam; all of us took photos of his shell collection. I attended to medical consultations with Marie (seizures) and her baby boy (dog bite), and afterward, we returned to the *Verona* with several of Norman's friends, including assistant district officer Mel and his wife, Shirley. Later that afternoon, we continued onward toward the Trobriand Islands.

By chance, I had heard of the Trobriand Islands several years earlier. In a college course, I had learned about Bronislaw Malinowski, a Polish-born anthropologist who spent several years living on these islands conducting ethnographic studies. His New Guinea sojourn occurred during World War I, when, after previously studying in London, his status as a subject of the Austro-Hungarian Empire prevented him from returning to England. One of his most famous books was the *Sexual Life of Savages in North-Western Melanesia: Marriage, and Family Life among the Natives of the Trobriand Islands, British New Guinea,* published in London in 1929. In addition to serving as a classic model of ethnographic research, in the Western world his book was received with fascination by academics but with dismay by a society in which frank and open discussion of sexual behavior was strictly taboo. Malinowski, who later became a giant among anthropologists, was in the United States when World War II broke out. He remained as a visiting professor at Yale University until his unexpected death in 1942.

The following quotes I selected from chapter 3 of his book to offer a sense of the cultural environment related to sexual behavior (italics mine):

> The Trobrianders are very free and easy in their sexual relations. To a superficial observer it might indeed appear that they are entirely untrammeled in these. This, however, is not the case; for their liberty has certain very well-defined limits. The best way of showing this will be to give a consecutive account of the various stages through which a man and a woman pass from childhood to maturity—a sort of sexual life-history of a representative couple.
>
> We shall have first to consider their earliest years, for *these natives begin their acquaintance with sex at a very tender age.* The unregulated and, as it were, capricious intercourse of these early years becomes systematized in adolescence into more or less stable intrigues, which later on develop into permanent liaisons. Connected with these latter stages of sexual life, there exists in the Trobriand Islands an extremely interesting institution, the bachelors' and unmarried girls' house, called by the natives "bukumatula"; it is of considerable importance, as it is one of those arrangements sanctioned by custom which might appear on the surface to be a form of "group-marriage"...
>
> The child's freedom and independence extend also to sexual matters. To begin with, *children hear of and witness much in the sexual life of their elders.* Within the house, where the parents have no possibility of finding privacy, a child has opportunities of acquiring practical information

concerning the sexual act. I was told that *no special precautions* are taken to prevent children from witnessing their parents' sexual enjoyment...

There are plenty of opportunities for both boys and girls to receive instruction in erotic matters from their companions. The children initiate each other into the mysteries of sexual life in a directly practical matter at a very early age...

The attitude of the grown-ups and even of the parents towards such infantile indulgence is either that of complete indifference or of complacency—they find it natural, and do not see why they should scold or interfere. Usually they show a kind of tolerant and amused interest, and discuss the love affairs of their children with easy jocularity...

...If we place the beginning of real sexual life at the age of six to eight in the case of girls, and ten to twelve in the case of boys, we shall probably not be erring very greatly in either direction...

...The little ones sometimes play, for instance, at house-building, and at family life. A small hut of sticks and boughs is constructed in a secluded part of the jungle, and a couple or more repair thither and play at husband and wife, prepare food and carry out or imitate as best they can the act of sex...

It is important to note that there is no interference by older persons in the sexual life of children. On rare occasions some old *man or woman* is suspected of taking a strong sexual interest in the children, and even of having intercourse with some of them. But I never found such suspicions supported even by a general consensus of opinion, and it was always considered both *improper and silly* for an older man or woman to have sexual dealings with a child.[1]

1. Bronislaw Malinowski, *The Sexual Life of Savages in North-Western Melanesia: An Ethnographic Account of Courtship, Marriage and Family Life among the Natives of the Trobriand Islands*, British New Guinea (Mansfield Centre, CT: Martino, 2012), 44-51.

Vic was clearly in his element in the Trobriands. On November 12, we anchored off of Iwa Island, Vic's recommendation and a place he obviously knew well. As several of us from the *Verona* walked to a village with Vic, it was immediately apparent that he was recognized and welcomed as he chatted with villagers in their native language. With camera in hand, I was in a photographer's paradise! And I even had with me a small tape (cassette) recorder, which I used to amaze locals as they heard their voices emerge from my magic box. While aboard the *Verona*, I had recorded several of Vic's stories, but his heavy accent made them a challenge to understand.

Figure 32. Vic Busuttin with Trobriand islanders, Iwa Island. Photo by author.

The simple houses of the neat village were spread out, and I followed Vic around as he made his rounds chatting with adults he knew, presumably with an eye for future "guests." Vic had told me that parents of his "guests" always approve of sending their daughters to his home, but that he does reward them with tobacco sticks and other gifts. At one home, Vic told me that he was going to take a "nap," so while he was napping, I returned to the beach, swam, and took photographs of a sailing outrigger that had just arrived from Kitava Island. It was constructed entirely from natural materials. Remarkably, these vessels and their mariners are capable of navigating between islands hundreds of miles apart.

At seven that evening, we departed for Kiriwina Island, arriving at 0800 the following morning and anchoring off the village of Kaibola. Tony, the village constable and Vic's friend, boarded the *Verona.* Later, I went with Vic and Tony to his house, and soon Vic left to do whatever he wanted. That afternoon I enjoyed superb scuba diving and spear fishing with Lloyd and Edgar and later photographed the seagoing sailing outrigger from Kitava.

Later that afternoon, while on the *Verona,* I was surprised to see a blond woman and a man arriving as passengers in a canoe paddled by an islander. They introduced themselves as Mary and John, he the education director in Losuia, a nearby village. After a tour of the *Verona,* I volunteered to take them to shore, where I also met Ian, the assistant district officer in Losuia, and his wife. Plans were made to meet them the next day for a trip to Losuia.

I returned to the *Verona* for the night, while Vic, grumbling, had agreed to take some of our crew on a crocodile hunt during the night. It seemed a big sacrifice for him because he likely had other plans. The hunt took place in a river, with flashlights an essential tool. Crocodile eyes apparently shine red, with the bright light leading to momentary paralysis and easy targeting.

In the morning, the *Verona* was surrounded by locals in canoes, some offering wares for sale, such as the grass skirts worn by all the women. In one beachfront area earlier that day, Trobriand females, clad only in grass skirts, could not understand why our women were not bare breasted like they were. Repeatedly, they attempted to remove the visitors' clothing to expose their breasts.

Figure 33. Iwa islanders fascinated by Western visitors. Photo by author.

About this time, we discovered that the crocodile hunters had returned from their overnight hunt with a nine-foot prize. Cameras captured Edgar's trophy while considerable man power was expended hauling it aboard.

Soon afterward, a jeep appeared on shore; Ian and his wife had arrived to drive us to Losuia. We stopped at several small villages on the way, observing with curiosity the colorful inhabitants and the finely crafted "houses" for storing their staple food, sweet potatoes. These storage facilities were elevated to protect the contents from vermin and seemed to be better constructed than the dwelling they accompanied. Finally, after a pleasant visit with Ian and his wife at their home, and meeting Tony (the doctor) and his wife, we returned to Kaibola village. Before boarding the *Verona*, I purchased an exquisitely carved ebony walking stick, a possession that now resides in the umbrella container in the entryway of our home.

Figure 34. Bill Bunting standing, Edgar Faust seated, admiring his crocodile. Photo by author.

Vic was still grumbling about his sleepless night hunting crocodiles when it was time for the *Verona* to set sail. I bade farewell to previously unimaginable Vic, a character who still makes my head spin when I think of him fifty-plus years later. The one letter I received from him nearly a year later was two pages long, written in a firm, clear hand, and it offered a newsy update. Now living near Losuia, on September 28, 1966, Vic wrote: "I am doing a job here for Public Works Department on a dredge. It is a very good job for me and I like it very much. My house is always crowded with the girls." Later, he asked me for advice: "I am not as good as I used to be Barney, perhaps you could advise me; may be old age. I am willing but *he* is not; only about twice a week now. They are turning up at my place faster than I can accommodate them."

Twelve

Peripatetic Dreaming III: Ceylon to Lisbon

March 7–11, 1966

The most striking facial features I observed during my year on the *Verona* were those of women and children of the Maldive Islands. This apparent blend of Arabic and Indian genes created faces with large, expressive eyes that reminded me of a popular artist of that era, Margaret Keane, famous for her "big eyes" paintings. The *Verona*'s arrival on March 7, 1966, at the capital, Malé, occurred decades before the Maldives reinvented themselves as a prime Indian Ocean jet-set resort destination and also long before this island nation of low-lying atolls and crystal-clear lagoons was recognized as facing a great national threat of destruction from global warming and rising ocean levels.

Two days after leaving Malé, near Horsburgh Atoll, we anchored for a workday. Afterward, the crew was free to engage in scuba diving in the atoll's magnificent lagoon or to otherwise relax. From the *Verona,* we could see a village, and to our surprise, what appeared to be Morse code light signals were directed toward us from shore. Curious, Chris and several of us took our motor launch to where a large group of men had assembled at the shoreline. It was then

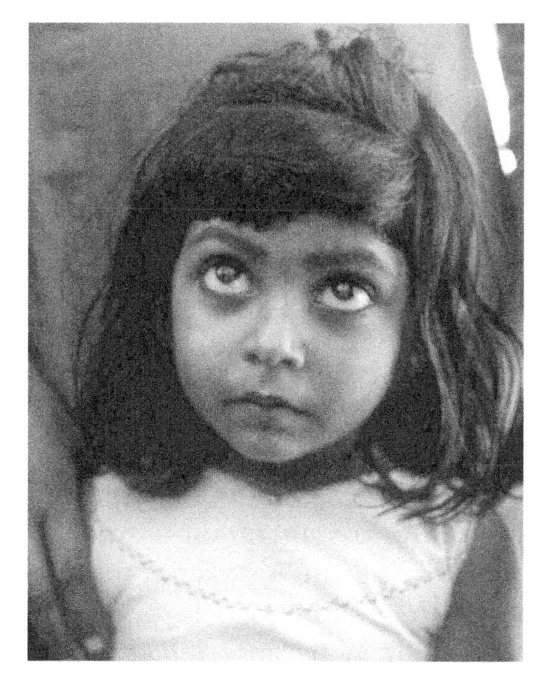

Figure 35. Young girl in Male, Maldive Islands, Indian Ocean. Photo by author.

we learned that we were at a prison island for political offenders. The prisoners were friendly, certainly not threatening toward us, and were curious to know why we were there. We then learned that the island was a strictly forbidden destination to visit without permission from the national authorities. Fortunately, we had not bumbled into a different, nearby prison island that was home to violent criminals.

April 15–19, 1966

For me, it was a joy, a gift, to experience Beirut and Lebanon for four glorious springtime days. Not many years later, the violent civil war (1975–90) that erupted in Lebanon denied this opportunity for foreign visitors. But in April 1966, Beirut was still known as the "Paris of the Mediterranean" and the "Lebanese Riviera." It was a city that offered a celebrated cuisine, casinos, snow skiing less than an hour's drive from welcoming beaches, and Christian and Muslim cultures peacefully intermingled—this all blended in a francophone ambience. Beirut responded to the needs and desires of sophisticated visitors from around the globe.

As a senior medical student on surgery rotation at Baylor College of Medicine, where Dr. Michael DeBakey (of Lebanese ancestry) reined as perhaps the world's most celebrated cardiovascular surgeon, I had come to know a Lebanese patient and the physician who accompanied him, Dr. Robert Hanna. The patient had come to Houston for heart surgery and remained in our training hospital for weeks of recuperation. I spoke with Dr. Hanna almost daily, and before he and his patient returned to Beirut, he insisted that I let him know if I ever were to visit Lebanon. I stayed in contact with Dr. Hanna because I had seriously considered applying for an internship at the American University of Beirut. Of course, this was some years before my *Verona* travels.

I was pleased to know someone in this bustling city. Once in Beirut, I called Dr. Hanna, and to my delight, he suggested we meet at the Riviera Hotel. After reconnecting with him over a drink, Dr. Hanna drove me to his home, where I met his wife and other family members. We then proceeded to a striking home in a nearby mountain village, where I met Emil Shehadi, a charming host and also owner of the Bliss Hotel in Beirut. Shehadi's friends, George F. Haddad and Dr. Mishalamy, were also present. Following a fine dinner and an introduction to the collection of exquisite antiques on the second floor of Emil Shehadi's home, George Haddad drove me to my hotel.

Emile Shehadi's interest in sailboats led to tour of the *Verona* the following day, and afterward again I was treated to lavish hospitality, including visits to Beirut's prized tourist destinations and dinner at a celebrated restaurant. That evening was my first exposure to Lebanon's famous meze, a vast array of more than twenty-four small dishes, hors d'oeuvres of the region, served with drinks. I was also introduced to arrack, an anise-flavored aperitif common throughout the Middle East and Mediterranean world. The Lebanese call arrack "lion's milk" because when mixed with water, it acquires a milky appearance.

The following day, April 17, I was treated by George Haddad to something quite extraordinary, a visit to the famous Roman ruins of Baalbek in the Bekaa Valley of Lebanon. Among the ruins stands what is said to be the largest Roman temple of antiquity. When civil war erupted in Lebanon in 1975, the Bekaa Valley was an epicenter of violence, and safe access to this celebrated site was denied for decades. I feel extremely fortunate to have seen it. We returned to Beirut that evening, and over dinner with George Haddad and Robert Hanna, I was immersed in conversation as I learned more about Lebanon's unique history, culture, and its relative wealth.

My penultimate day in Beirut was spent writing letters, shopping, purchasing a *Time* magazine, attending to photographic needs, and, with other crew members, absorbing more of the ambiance and tempting food items displayed at innumerable restaurants. During my final day, I again greatly enjoyed the company of my principal Lebanese host, Dr. Robert Hanna, over a leisurely lunch followed by the cinema (*The Sound of Music*), more tasty nibbles, lion's milk, and farewells. Near midnight, the *Verona* departed for the Greek island of Rhodes, the largest of the Dodecanese (twelve islands), not far from the coast of Turkey.

April 22–25, 1966

Log entry: *"Arrive Rhodos 16:00. Tie up stern-first in inner harbor with other yachts. Many tourists gawking at the Verona."*

The ancient Greeks had their version of a bucket list: the Seven Wonders of the Ancient World included sites that dedicated travelers must visit. One of them, the Colossus of Rhodes, may have stood at the entrance to the very harbor into which the *Verona* glided this sunny afternoon. History indicates that the Colossus was constructed nearly 300 years BCE and was destroyed by an earthquake some sixty years after it was completed. The statue is thought to have been approximately the size of the Statue of Liberty in New York Harbor. It depicted the sun god Helios.

Figure 36. Harbor of Rhodes, where the Colossus of Rhodes may have stood. Photo by author.

The Greek island of Rhodes was a wildly popular destination for Scandinavians, and the streets and waterfront of the principal city, Rhodes, were packed with warmth-seeking Swedes. April 23 was a workday on the *Verona,* but surrounded by curious spectators on this gorgeous day, the task of scrubbing the topside was completed quickly, and we were rewarded with an evening stroll around the ancient walled city. Despite the warm reception we had received at earlier ports of call, this first exposure to Greek islanders expanded my concept of hospitality. It seemed to me, at least in part, the warmth I experienced was related to my American citizenship. Upon learning my nationality, the faces of those who spoke English lit up and they proceeded to talk excitedly about their time living in America or about their relatives in the United States. Most of the locals I met were of modest means, but their hospitality seemed to explode from every direction. This was but a prelude to experiences on other Greek islands.

Rented Vespas provided the means of exploring the island with a few crewmates. As we headed toward the ancient town of Lindos on April 24, we saw oceans of blooming wildflowers, tiny white villages nestled on hilltops, and smiling farmers in fields and on their donkeys. Soon Lindos and the Temple of Athena on the mountaintop appeared. To avoid the crush of tourists at the temple, we decided to skip visiting it. Instead, we bought sardines and bread and cheese in a tiny store and rode on. We soon stopped near the village of Lardos to enjoy a picnic in a magnificent meadow of blooming poppies and dandelions. After lunch, we continued circling Rhodes's northern half as we rode through Laerma and Embona. In the latter village, we refreshed ourselves by sampling local wine and dolmas (stuffed grape leaves). We then passed through Kamiros on the west coast and arrived back in Rhodes town before dark. After the long day, I was almost too exhausted to partake of the ongoing good cheer surrounding the *Verona*'s berth on the harbor.

Because Lindos was such a popular destination on Rhodes, the *Verona* headed there before sunrise the following day and arrived by noon. Most of us who had already been there remained aboard to enjoy the view and the silence while the others left the *Verona* to explore Lindos. In the late afternoon, three of us decided to visit Saint Paul's Bay, just south of Lindos. From our launch, we viewed the rocky bay in which Saint Paul had sought refuge during a storm, and we entered the narrow passage leading into the well-protected bay. Entering this narrow passage would have been a treacherous feat during a storm. We saw a small chapel built to commemorate Paul's shelter here and also were rewarded with a view of the sun's late afternoon rays on the pillars of the Temple of Athena in Lindos. The *Verona* departed later that evening for Crete or Santorini, depending upon the whims of the wind gods.

April 26–27, 1966

Log entry: *"Destination still undecided. Chris decides to power to Astympalia, anchoring about 20:00 in bay north of village."*

The next day was spent roaming around this small, friendly, infrequently visited island in the Aegean Sea. To my surprise, about noon, I found Chris in a taverna sipping ouzo and solo dancing to Greek music, a common practice by Greek men in this part of the world. I joined him for a lunch of grilled just-out-of-the-ocean fish. At 18:30, the *Verona* weighed anchor, our next port of call uncertain.

April 28, 1966

With pumice floating in the water around us to remind us of the volcanic nature of Santorini, the *Verona* tied up to a buoy below the town of Firá. From the tiny harbor, hundreds of steps led up the steep cliff, part of the caldera of an ancient volcano. The *Verona* remained at Santorini for two days.

May 31, 1966

The *Verona* arrived in Lisbon—terminus of the year-one voyage.

But for me, a lifetime connected to Santorini and the village of Oia had begun.

Thirteen

My Village, Oia

My village, Oia, perched high over the Aegean Sea at the northern tip of Santorini, remains a place of astounding beauty despite having evolved during my lifetime from grinding poverty into one of the Mediterranean's most coveted destinations. *Webster's* definition of *surrealistic*, "having a strange dreamlike atmosphere or quality," aptly captures today's Oia as much as it did when I first arrived in April 1966. Now the ruins of its traditional cave houses have been restored into freshly whitewashed dwellings, and gleaming boutique hotels charge up to four figures per night. Across the landscape, windmills stand at attention, flaunting their full form against a vast curtain of blue sky. A former shoemaker's shop is now an elegant restaurant, and other family-operated enterprises now serve tourists as high-end jewelry shops, art galleries, and other emporia catering to the ultrarich. Many of these elegant shops are operated during the peak tourist season by Athenian entrepreneurs.

Deep-blue Aegean Sea fills what remains of the ten-mile diameter crater that was formed when the sky was darkened worldwide 3,700 years ago by the most forceful volcanic eruption of recorded history. Following others through the centuries, a violent 1956 earthquake virtually destroyed Oia, but a few homes of once-prosperous sea captains and ship owners remained standing. Today, only a few pockets of ruins are visible, reminders of families that vanished or living relatives who could not be located.

Today's airport on Santorini opened in 1972 to scant air traffic. Now it receives commercial and charter flights from nearly every major city in Europe and some from Asia. In recent years, photos of "destination weddings" in Oia, especially for Asian couples, have become powerful status symbols in home countries (Japan, China, South Korea, and Taiwan). At any time of the day, but mostly early morning or at sunset, an astounding number of Asian brides in flowing wedding gowns scramble over patios, rooftops, and walls to strike dramatic poses for their trailing photographers and spouses. Yes, surrealistic, this frenzied burst of cross-cultural performance art on an Aegean island. Also surreal is the sight of caravans of monstrous modern buses bearing hundreds of cruise-ship vacationers preparing to clog the narrow alleys of Oia in the late afternoon. This is a typical summertime day on an island that in 1966 boasted five taxis and six trucks, all Soviet-manufactured Ladas.

The simple "cave house" in Oia that Nancy and I lovingly restored and owned for some thirty-five years now belongs to Markos Karvounis, the son of our original caretakers, Teodosis and Youlia Karvounis. Markos's older brothers, Adonis (Neptune Restaurant) and Manolis (Sunset Bar), round out the core family. While we no longer own the house, the times we and our children spent in Oia have permitted us to acquire a possession of incalculable value, our beloved "Greek family." Markos, whom I first met when he was six, then helped his parents by waiting on tables in their tiny tavern. Markos purchased our house in 2006, and his promise then to provide housing for us during return trips has been fulfilled repeatedly despite our protests. Our son and his bride spent their honeymoon in Oia more than a decade ago. Markos now has a talented and beautiful wife, Torun, and son, Rafael. We attended and helped celebrate their unforgettable wedding in Oia, and two years later we returned to meet baby Rafael. Although Youlia survived to attend Markos and Torun's wedding, both she and Theodosis are now deceased. Markos, as the youngest son, meticulously maintains his parents' graves in the village cemetery.

Now back to my discovery of Oia. Morning, April 28, 1966. The evening of the previous day, we bid farewell to the tiny Aegean island of Astypalaia and sailed in a westerly direction. French ketch the *Europa*, which we had seen in Rhodes, arrived just before we weighed anchor, still uncertain of our next port of call. Depending upon wind direction, our options were Firá, Santorini, or Heraklion, Crete. Overnight, the waves had been pounding rhythmically against the teak hull just inches from my head, and the roller-coaster motion gently pressed and lifted my body, but minutes before my 0400 watch duty was to begin, a sensation of smooth gliding abruptly replaced the pounding.

Rolling from my bunk and climbing the few steps to the deck, a dense morning mist added to my confusion. "What's going on? Why is it so calm?" I asked the first crew member I encountered on deck. I learned that during the night, the wind had shifted such that Santorini, the towering crescent-shaped remains of ancient volcanic eruptions, became our destination. Years later, after this wildly beautiful island had profoundly enriched my life, the answer to the question "Why did you go there?" was simple: the wind blew me there.

Why the abrupt silence? As the *Verona* entered the ancient Aegean-filled volcanic crater, the wind was extinguished by the towering cliffs of the caldera. As I looked up through the mist, I could see ghostly clusters of ruins, cave houses, I later learned, destroyed by the 1956 earthquake. Oia, clinging tenuously to the top of the caldera's almost-vertical cliffs at the northernmost tip of Santorini, had suffered profound damage.

While a medical student, a few years before the *Verona* trip, I had been captivated by *The Story of San Michele*, the autobiography of Swedish physician Axel Munthe. Munthe, as a young medical student in Paris in the late nineteenth century, had briefly traveled by boat to the island of Capri and to the ruin-filled mountainside village of Anacapri. In a fit of youthful passion, he committed to return to build a home from the ruins of Emperor Nero's villa. Munthe ultimately realized his dream. San Michele, his Anacapri home, became a defining element of his long, adventuresome, inspired life. While I never consciously attempted to emulate Munthe, in retrospect, years later, the parallels became apparent to me. Subconsciously, perhaps Oia was my Anacapri.

In addition to my immediate fascination with Santorini, I had an intense longing of which I had become aware after a few months at sea: a burning desire for a place where I could remain beyond the usual one to two days. Repeatedly, I was frustrated by the painful and obligatory need to part with new friends and acquaintances. I longed to find a port of call where I could remain (with permission) for a week or more.

The *Verona* continued its glide past Oia until it reached the cliffs below Firá, Santorini's main town. We tied up to a buoy near the tiny harbor from which ascended mule-trodden stone steps leading to the town hundreds of feet above. About 07:00 the sun was barely above the horizon, hidden from view by the cliffs, so at this early hour I was surprised to see three men standing near the waterfront, deeply engaged in conversation. They were obviously not fishermen, and I wondered what they were doing. I jumped into our launch to be one the first crew members on shore, landing near the men. They glanced at the *Verona*, perhaps curious to learn more about it and its crew. Peter Klap, a Dutchman and the most outgoing of the three, engaged me in conversation. He listened as I briefly described our crew and itinerary, and I learned that Peter and his wife, Rose-Marie, lived in Zurich but had a house in Firá and spent long periods on the island. His companions at the waterfront were John Kunzler, a Swissair steward who likewise lived in Switzerland but visited Santorini often, and Oscar, a Frenchman. Years later, I discovered what took them to the waterfront so early that morning: they were looking at unoccupied, earthquake-damaged ruins that they might purchase and restore.

Already dazzled by the vistas surrounding me, and reassured by the welcoming reception I had experienced, I was eager to explore Firá and beyond. Peter, older than I by a decade or two, had a sparkling, appealing manner, and as we parted he said, "Come to my house this afternoon for a drink." Delighted to do so, I asked how to find his house. Peter smiled and chuckled and said, "Anyone on the street can direct you to Spiti Petros [Peter's House]." I doubted it could be that simple, but it was. Late that afternoon I explored the picturesque streets of Firá and gazed with amazement at the caldera and its enormous crater and the secondary volcano at its center. The first young boy I asked "*Pou ine Spiti Petros?*" ("Where is Peter's House?") led me there in minutes. I had by now learned a few words of Greek.

Multilingual Peter offered a splendid introduction to Santorini because he had lived there on and off for years, was fluent in Greek, was knowledgeable about local customs, and seemed to be a friend of virtually everyone. During the 1960s, much of Greece remained desperately poor, a "third-world" country. It suffered massively during World War II as the Greeks vainly attempted to resist occupation by Italian and later German military forces. Starvation killed many. Not long after the war ended, Greece experienced a soul-wrenching and horrendous civil war, the scars of which linger today. The powerful nonfiction book *Eleni,* by Nicolas

Gage, offers readers a vivid view of this painful chapter in Greek life, and *Captain Corelli's Mandolin,* by Louis de Bernières, serves up a lighter, history-based picture of the World War II years on the Greek island of Cephalonia. Except for a few tourist destinations, the Greek islands, isolated and without dependable transport, shared the poverty and endless political strife of the entire nation.

As I explored Santorini, I found that Americans were warmly and sincerely welcomed. Almost without fail, one or more of the older men gathered at village coffeehouses, fondling their worry beads, had lived and worked in Cleveland or New York or another major city before returning to their home village to retire. Despite the "first-world" influence of these aged Greek Americans, many of whom returned to marry young village women, the island's culture remained quite intact and profoundly traditional. I was astounded to learn that following meals and drinking in a village tavern, it was considered inappropriate for locals to ask for the bill. Rather, the tavern owner knew that someone would return in a day or two to settle the account, at which time the owner would casually tear off a piece of scrap paper, scribble a few numbers and present the "bill." In those days, it rarely exceeded one dollar per patron.

Peter's early life in the Netherlands as part of the Dutch resistance to the Nazi occupation taught him to be resourceful. He came from a family of artists and was a man of many talents and passions, one of which (in addition to being a concert pianist and painter) was restoring ruined houses. From Peter's perspective, the ruined cave houses were like candy to a kid, abundant and inexpensive. Many families had relocated to the mainland and ruins could be purchased for a pittance. At this time, economic opportunities on Santorini were severely limited and mostly agricultural, such as growing tomatoes on small farms or cultivating small vineyards. Tourism was limited to periodic cruise ships, which visited long enough for adventuresome visitors to climb or ride a mule up the 495 steps to Firá. Mules were the main source of transport in those days; the few taxis and trucks were the only motorized vehicles on the island. The locals firmly believed that the mules contained the souls of sinners in purgatory and consequently deserved to be treated badly.

How could I foresee during my first evening at Spiti Petros that Peter and his charming Swiss-Italian wife, Rose-Marie, would become almost family, or that Peter would spend countless hours helping me acquire a ruin in Oia? How could I have known that during the decades to follow, we would repeatedly exchange visits in the United States, Zurich, and Santorini, or that Peter and Rose-Marie would be with us at our home in Texas just weeks before Peter died of lung cancer?

I was so taken by the magic of this island, and with Peter's hospitality and offer of a room, that I was convinced that Santorini was the place I had been longing for. Finally, I hoped, I could savor newly created friendships without sailing off in a day or two. Anxiously, I asked Chris, and happily he agreed, perhaps in part because I had introduced him to Peter, and that evening, we had all been invited to Spiti Petros. Chris's brother John from New York had joined the *Verona* in Greece, and he too partook of Dionysian excesses while sitting on Peter's terrace with wine glasses reflecting the sun's slow descent into the Aegean.

Staying at Peter's home introduced me to the circle of close friends of this exuberant Dutchman. I met Nikos Kafieris, a highly respected housebuilder, and his wife. I also met Giorgos Skopelitis, son of a Greek Orthodox priest in Firá, who assisted Nikos. Several late afternoons, the four of us ended up in nearby Firostefani, in tiny Taverna Russos owned by warmhearted and outgoing Giorgos Russos and his wife, Argarula. Although we had not discussed it, Peter seemed aware that my irrational mind was already entertaining the prospect of owning a simple peasant cottage perched over the sea. For this reason, he had arranged for me to meet and spend time with Nikos. I soon learned that Peter had his eye on tranquil Oia as a place for a second Santorini home, far removed from the bustle of Firá, which could be crowded and noisy during July and August.

Having first viewed Oia through the morning mist clinging to towering cliffs, the thought of owning a house there became fixed in my brain. By now the *Verona* had left, and I was excited about walking the streets of Oia. Peter was perhaps dreaming also of a second house as the tiny creaking bus transported us the seven miles between Firá and Oia. The view from the bus window would induce terror for any passenger fearful of heights: the narrow unpaved road had no protective railing, and the drop-off was hundreds of feet. Arriving, I found Oia equally enchanting from above as from below. The views of the caldera were even more breathtaking than from Firá, and the tiny silent village perched along the ridge seemed ghostlike because of the whitewashed houses, obvious poverty, scarcity of inhabitants, absence of electricity, and most of all, because of hundreds of pockmarked remains of earthquake-destroyed traditional cave homes. The scarcity of timber on the island, plus the relative ease of digging into pumice-like volcanic stone, contributed to the unique architecture and home construction in Oia. Straight lines were a rarity, but plaster and whitewash applied by talented craftsmen transformed the humble caves into simple but spectacular homes and terraces.

Figure 37. Oia, Santorini, Greece in 1967, showing pockmarked ruins of cave houses. Photo by author.

Peter's network of local friends and acquaintances identified three ruins for me to visit. He explained that obtaining legal title to property was complicated by the need to locate the owner or owners and that many of the surviving families had scattered to the mainland or to other islands. All three of the ruins we visited appealed to me. Each offered a terrace with a glorious view of the caldera.

During lunch in Oia's sole tavern, the Karvounises' youngest son, Markos, about six and exuding a confident and observant air, wandered in. He smiled at us non-Greeks before entering the tiny kitchen to greet his mother. Marcos's oldest brother, Adonis, was employed as a cook on a Greek merchant ship, and his other brother, Manolis, several years older than Markos, was still a schoolboy on Santorini. This was my introduction to the Karvounis family. After lunch, we visited the ruin that Peter was destined to acquire at the northern end of Oia.

The week I spent at Peter and Rose-Marie's house in Firá instilled a passionate attraction to this rock-strewn Aegean island. In addition to being astonished by the endless surreal vistas, my appreciation of the local townspeople grew. Rarely did they hesitate to share their meager possessions with strangers. With Peter and a few of his friends, I attended a local religious celebration in the village of Pirgos. Gathered in a dimly lit room filled with the sounds of local musicians, we partook of the feast consisting of bread, soup, and canned sardines—and of course, ample red wine produced locally.

Another special moment occurred when Peter and I and Nikos and his wife gathered one evening at Taverna Roussos. Of course, libations appeared at the table to celebrate the moment. After a while, Nikos's wife excused herself to go home and prepare dinner. A second round of libations then appeared. Before long, traditional Greek bouzouki music filled the room, and one of Giorgos's sons, about age five, started dancing. Soon Peter and Giorgos Roussos joined him in a vigorous island dance. Spirits were high and getting higher when several of the group wandered into the taverna's cramped kitchen, where Giorgos's wife, Argarula, was preparing the evening's culinary offerings, including a very large bowl of shredded cabbage (*lachano* in Greek). One of the men was moved by the joyous moment to grab a handful of cabbage and toss it into the air like confetti. This Greek-like celebration of free will continued until everyone had participated, the bowl had been emptied, and heads and all surfaces were covered with bits of cabbage. To this day, the famous "*lachano* party" is part of the Roussos family collective memory, and we laugh about it when we are together.

Figure 38. Above: Taverna Roussos, Firostefani, Santorini in 1966, with Argarula on terrace; Below: Nikos Kafieris (left) with Peter Klap (right) before the "*lachano* party". Photos by author.

I left Santorini after a week or so, painfully bidding farewell to new friends. As I traveled by ferry and overland to meet the

Verona in Lisbon, my head was filled with thoughts of returning soon to claim a ruin in Oia for my very own. The *Verona*'s first-year voyage was scheduled to end in Lisbon, where it would soon be boarded by a new crew. It was time, of course, for me to return to the United States, visit my parents, and later move to Baltimore for postdoctoral studies.

After saying farewell to my *Verona* crewmates in Lisbon and returning to the United States, the ongoing exchange of letters with Peter kept me anxiously awaiting mail delivery at my parents' home in San Antonio and later at my new residence in Baltimore, where I entered a training program at Johns Hopkins School of Hygiene and Public Health. I still have seven letters received from Peter during 1966 and 1967. Whether the postmark was from Santorini or Zurich or Amsterdam, Peter was a faithful correspondent. The early letters kept me posted on complicated, failed house-purchasing efforts. For example, Peter and Nikos and Giorgos Skopelitis, acting on my behalf, had invested countless hours searching for and cajoling family members of one of the three houses I had visited initially. It involved two trips from Santorini to Athens by Giorgos to find family members and obtain permission, but ultimately one of three sisters refused to sell. The demand for ruins on Santorini was increasing, as were prices, and an architect from Athens had purchased more than ten ruins in Oia. After anxious months, Peter's exuberant letter (written on Olympic Airlines stationery as he flew to Zurich from Athens) finally arrived. It started:

> My dear friend Barney, My goodness! At LAST I am able to tell you, I bought a house for you in Oia. I will not tell you how difficult it was to find it, how many taxis to Oia it took, how many disappointments we had to face, we all, Giorgos [Skopelitis], Nikos, Argarula, Giorgos Roussos, and all other friends were convinced of one thing: we had to find

> a house for Barney! The house you wanted was finally *not* for sale.

And at the bottom of the final page, Peter wrote: "My goodness! Barney now dance and take a drink to my health. I know I deserve it! I send you warm greetings from all our friends on Santorini, and please write back soon to your old friend Petros Klap."

I also learned that Greek law controlling the sale of property on Santorini to non-Greek citizens was on the verge of changing. But I did not know that at the closing in August 1967, I would be the *last* foreigner to receive an official deed to property on Santorini. Recent and dramatic archeological discoveries on Santorini had precipitated the need to declare it a protected archeological zone. Also, elevated tensions with archenemy Turkey had led to additional legislation declaring Santorini to lie within a "security zone" that excluded non-Greek ownership of property. While this may seem a modest legal technicality, tales abound of foreigners who built or restored homes in the name of Greek citizen "friends" who later found themselves with no property whatsoever. And yes, the cost of my ruin was about $500, including all the land down to the sea and taxes exempted in perpetuity, plus about $150 for travel to Athens and related expenses.

Despite US efforts to stabilize Greece after World War II, the country reeled from one crisis to another, leading in April 1967 to the seven-year "regime of the colonels," a dictatorship. So when I returned briefly in August 1967 to transfer Peter's ownership into my name, the inhabitants of Santorini were adjusting to a new set of rules enforced by a local strongman appointed by the colonels. For example, the colonels had outlawed the centuries-long cathartic tradition of breaking plates and glasses as a form of exuberant celebration. They apparently considered it primitive behavior. But in a taverna in Firá, after I was handed a deed to property destined to become Spiti Iatros (House of the Doctor), I crossed a very meaningful personal threshold. Along with Peter and Nikos and others, I could be seen smashing plates and glasses on the floor that evening after toasting to my new ruin. Nobody seemed particularly concerned about the colonels' edict. I learned that on Santorini celebrations were commonly rated by the cost of breakage. As I recall, on returning to the taverna the following day, I was presented with a breakage bill of about thirty dollars, miniscule compared to epic parties I heard about.

Why do Greeks break dishes? I have pondered this question, and have conjured up three main reasons—my opinion only. First, the act says that material things are unimportant relative to the joy or emotion experienced. Second, the act of shattering an object freezes a moment in time—that glass or plate will never again share such a moment. And third, it is an expression of humankind's unfettered freedom, a very Greek trait. And yes, it was a cathartic experience for me.

Peter and Nikos advised that the first step in restoring Spiti Iatros was to rebuild the cistern and rainwater-collection system. Water was precious in Oia, the only source then coming from the sky, and that quantity was quite modest. For this reason, every house was designed as a rainwater-collection device, using gravity to direct the precious commodity into a cistern. For Spiti Iatros, including the restoration of a large terrace and wall, this work cost more than the initial purchase price of the ruin and was completed by Nikos within a couple of years.

Nancy and I were married on May 16, 1970, and of course Santorini was the principal destination on our honeymoon a month later. Expecting to meet with Nikos to plan the restoration, we were aided by sketches executed by Peter. When we reached Santorini, we learned that a kidney stone had required Nikos to unexpectedly seek medical care in Athens and that his brother would represent him. We vividly recall the hours spent with Nikos's very conscientious brother. The English-Greek dictionary I clutched helped save the day, and we were assisted by liberal use of nonverbal communication. For example, the door separating the dining room from the guest room was so low that I almost had to go to my knees to pass through. Asked how big the door should be, I stood tall and reached my arm upward as far as possible and making a sweeping motion, said, "This big," scratching the cave wall with a nail. In the small area designated to become a kitchen, we were asked whether we wanted the sink to be of stainless steel or marble. Learning that stainless-steel sinks were more expensive than marble (because they were imported and considered a status symbol), Nancy and I did not hesitate to respond that we would "make do with marble." Remember: the main street of Oia is paved with marble.

Figure 39. Nancy at front door of *Spiti Iatros,* on our 1970 "honeymoon visit" after cistern and terrace repair was completed. Photo by author.

Figure 40. Spiti Iatros on our arrival the summer of 1971, before we painted the doors and windows blue. Photo by author.

After completing my studies at Johns Hopkins and at the University of California, Berkeley, I accepted a position with the Centers for Disease Control (CDC), Atlanta, at a tropical-disease research laboratory in San Juan, Puerto Rico. Now with a steady income, we could ask Nikos to continue with the planned restoration of our Greek ruin. So our first year of married life in Puerto Rico included the additional anticipation of organizing a return to Oia in the summer of 1971 to see Nikos's restoration. The gods smiled upon us when Nancy met Kate Trotman, via the Spanish tutor they shared. Kate and her husband, Art, were our neighbors in Old San Juan. The Clines and Trotmans bonded quickly, remaining bosom buddies and frequent travel companions to this day.

We described to Art and Kate our plans to occupy and furnish Spiti Iatros, hoping that this adventuresome couple would take the bait and join us. They did. Whenever together we smile when recalling improbable shared experiences. An example is the feat of transporting a refrigerator to the ferry that would take us to Oia from the Greek mainland. The scenario: the appliance store in Piraeus, several blocks from the pier, opened at 0800, and the day's only ferry would depart thirty minutes later. The plan: Kate would wait at the ferry with all our luggage and delay boarding until we appeared with the refrigerator. Nancy and I would purchase the refrigerator the moment the store opened and then, with Art's help, rush it to the pier. Purchasing the medium-sized Frigidaire went surprisingly quickly. We carried it to the street only to realize no taxis could be found that early. Now we had fewer than fifteen minutes until the ferry departed. The store owner saved the moment by producing a cart of sorts with very small wheels, and thus began a comical frantic dash to the pier over mostly cobblestone streets. With shouting in various languages, strangers helping push the wobbly cart and keep the fridge upright, and others leading interference on the streets, we miraculously arrived seconds before the scheduled departure. The compassionate Greek crew took pity on us comical foreigners and delayed departure for a few minutes while our possessions were moved aboard and tickets presented. That Frigidaire ultimately reached Spiti Iatros on the back of a mule and was still operating some thirty years later.

Within a few days, with substantial sweat equity from the Clines and Trotmans, Spiti Iatros was transformed into a home. All furnishings and decorations were acquired locally. Nancy's fondness of blue and white (colors of the Greek flag) and her innate sense of decor and love of decorating turned our cave house into a veritable love nest for almost-newlyweds. Our very basic but glistening white tile bathroom, with flush toilet and shower with hot water, seemed to be the talk of the neighborhood. Apparently, ours was one of the very few "modern" bathrooms in Oia. Neighbors who visited seemed far more interested in seeing it than our house.

To Nancy's credit (helped by thrifty local purchases), almost all who entered Spiti Iatros commented that it looked like a "Greek house," not like the homes of most foreigners.

As a lifelong foodie, I occupied myself with acquiring the basic needs of our kitchen and of exploring food markets and the like. Breakfast on the terrace with coffee, bread still warm from the local bakery, luscious yogurt, jams, and fruit became a ritual. My food-related vocabulary in Greek grew in step with our periodic visits to Oia's then-only tiny "supermarket," where feta cheese and other ingredients for a Greek salad were available. That became our typical lunch at home, and in the evening, we usually dined at Theodosios's tavern.

For decades, we somehow managed to return, often with our two children, every two to three years, but during our absence, we kept in touch with Peter and periodically with Theodosis, who became our caretaker. Visits required a note to Theodosis to advise him of our arrival date so that Youlia could air out the house and wash the linens. Every few years, a fresh coat of whitewash was in order. At the end of a stay, we would meet with Theodosis, who (after electric power reached Oia) would meticulously calculate the accumulated total of the electricity bill that he had paid for us and other minor expenses. He was always pleased with the modest extra amount we added for his time and effort.

These were the golden years for us, because life in Oia was the epitome of rural simplicity and tranquility in a spectacular Aegean village. We were pleased with the small but growing number of foreign visitors and homeowners, mostly German and Dutch, who contributed diversity and local economic benefit. Although the local economy slowly improved, Oia remained a poor village that was certainly off the radar for most travelers.

These years overflow with sweet memories. Our son, Philip, age five in 1979, insisted that he could find the bakery by himself and, armed with coins and the Greek word for bread, headed down the marble-paved street in the right direction. Soon, his chest puffed up with pride, he returned clutching two warm loaves. His two-year-old sister, Lea, observed all this and announced, "Daddy, when I'm big *I* am going to get the bread." For the next few years, they shared this prized chore. And there was the visit when young Lea refused to drink the ultra-pasteurized milk purchased in Oia but on return home, with a glass of cold milk in hand, declared, "Daddy, I'm so glad to be back to *our cows*!" Years later, I recall, Philip was old enough to rent a moped, and of course his favorite destination was what had become the "topless beach" on the other end of island.

With mixed emotions, we have witnessed the dazzling transformation of Oia from relative obscurity into prized global destination. On our most recent travels to Oia and the Karvounis family in June 2017, we learned that commercial aircraft now land on Santorini at a rate of about 325 per week. The facility is grossly inadequate, but an expanded terminal is under construction. Heavy tourism now starts in April and extends into November.

Despite this transformation, Oia fully deserves its distinction of being the most desirable village on the island, and Santorini the most remarkable destination in the Aegean Sea.

Now more than fifty years since my first view of earthquake-shattered Oia through the early morning mist, I reflect upon the *Verona*'s most meaningful gift, and I thank the wind that blew me there.

Postscript

As a neophyte author, I was startled by the degree to which writing a memoir alters one's sense of self. For example, my logbook notes revealed to me that I have *always* been a foodie. I had assumed that the pleasure I derive from experimenting in the kitchen and from exploring new cuisines was a trait acquired much later in life as "exotic" foodstuffs became commonplace in the United States. Similarly, although I characteristically enjoyed meeting strangers and making new acquaintances, reading my log after many decades, I was struck by the degree of social intercourse I engaged in when the *Verona* delivered me to a new port of call. It seems that I was almost always one of the first to disembark, eager to embrace a new destination.

Reviewing old photographs and related notes, reading forgotten letters, listening to taped exchanges with my long-deceased parents, and writing became a vehicle to revisit a remote time of my life and to learn about who I was then and how that differs from who I am now. A valued friend posed an astute challenge by asking *how* the *Verona* voyage changed me. He wanted to know in what ways I had evolved differently. I take his challenge seriously, but it is not simple to separate how the voyage changed me from how the *process* of creating this book changed me.

Like John Steinbeck in his *The Log from the Sea of Cortez*, my feeling of connectedness to the saltwater from which mankind evolved was greatly enhanced by floating on and in it for a year. During countless hours on deck, especially on watch at night, I sensed a haunting connection with mariners and with all seafaring people who through the millennia plunged into an unknown world and faced fears and dangers far beyond those that we usually face. I learned that to travel the world as a physician facilitates crossing cultural barriers and connecting with strangers, but the facilitating skill could just as easily be musical or linguistic or artistic. René Descarte said this about travel: "Travel is almost like talking with men of other centuries," suggesting that travel moves us through dimensions of space *and* time. And yes, looking back, I realize that my relative youth and unmarried status helped minimize barriers to exploring the new worlds to which the *Verona* delivered me.

I do believe that without the *Verona* experience I would not have gravitated toward a career in academia and tropical medicine, a field for which the world literally became my "laboratory." *Webster's* defines *cosmopolitan* as "not bound by national habits or prejudices; at home in all countries and places." My year on the *Verona* did not just expand my world. It exploded it. My awareness of and fascination with the astounding diversity of humankind was reinforced on a daily basis for a year. How can I not be thankful for that fortuitous gift?

Chris, whose many positive qualities I never ceased to admire, contributed to my sense of comfort and purpose on the *Verona*, and we spent good times together dining onshore or on side trips, such as to Taipei and to the Taroko Gorge in Taiwan. My friendship with Chris was meaningful for me and, I believe, for him as well. But at the end of the year, I was not certain I knew Chris any better than before. While I accept that my youth then hindered my understanding of his complex nature, even today Chris remains an enigma.

Chapter 2

Lloyd Davidson

Lloyd continues to contribute to the culture and ecology of Honduras by reintroducing macaws into their native habitats from which they had been eliminated. Lloyd still lives in Copán Ruins, Honduras, near Macaw Mountain and not far from Cafe Miramundo. Michele Braun remains on Roatán, Honduras, and the two get together often. Nancy and I visited them in Roatán in 2015; both were thriving with very active, productive lives, intersecting periodically.

Chapter 5

Pitcairn Island

I understand that the island has become far more accessible to tourists. Arriving by air is not possible, but one can fly to Mangareva (the nearest airport) some 330 miles away. From there, a ship visits Pitcairn about every three months. An exceptional source of information about Pitcairn is found at the Pitcairn Islands Study Center at Pacific Union College, California.

Sam Reese Sheppard

Despite my repeated efforts over the years, I have not succeeded in making contact with this good man. His life has been dedicated to exonerating his father's initial conviction.

Palmerston Atoll

I remain in e-mail contact with Dr. Joe Williams, who lives in Auckland, New Zealand. Thus far, I have not revisited Palmerston Atoll with him.

Vanuatu

I believe that the John Frum movement persists, along with other "cargo cults" in parts of Melanesia.

Chapter 6

North Sentinel Island

Thankfully, the inhabitants of this island remain "the most isolated tribe on earth."

Chapter 7

Yemen

Sadly, Yemen is in the grips of a horrific civil war, largely fought as a proxy by regional powers.

Chapter 8

Taiwan

Candice Bergen remains one of the most respected women in Hollywood. A friend of ours who knows her conveyed a few photographs I had taken of her in 1965.

Angkor Wat

This temple complex near Siem Reap, Cambodia, has become one of the world's most popular destinations for tourists.

Chapter 9

Stellan Moerner

Treasure hunters continue to seek the lost treasure of Atahuallpa.

Chapter 10

Miles Barne

Nancy and I maintain active contact with Miles and Tessa Barne. Miles has been pleased to turn over the management of Sotterley Estate to his son, Thomas. We expect to see Miles and Tessa in 2018.

Chapter 12

Maldive Islands

In addition to the massive devastation inflicted by a 2004 tsunami, this island nation in the Indian Ocean has also faced extreme political turmoil in recent years. Furthermore, rising sea levels secondary to global warming continue to threaten its national survival.

Chapter 13

Oia, Santorini

Now an immensely popular village on a favored Aegean island, Oia retains its striking beauty despite the transformation it has undergone in recent decades. In 2018, Markos Karvounis expects to complete his breathtaking new boutique hotel, constructed below former Spiti Iatros on the property he purchased from us. In June 2017, Nancy and I enjoyed a lovely visit with Markos and his family in Oia. Our love affair with Oia continues.

Acknowledgments

Nancy, my wife of forty-seven years, merits immense appreciation not only for being a partner in my life story, but also for her patience, support, unfailingly sound advice, and splendid editing skills. I could not have completed my book without her.

Our daughter, Lea Kimberly Cline, a university professor and accomplished academician, saved countless tedious hours by helping me finalize the formatting and the footnotes of the manuscript. Gerald Cline, my brother and a passionate world traveler and lover of maps, generously offered to prepare the images showing the *Verona*'s oceanic routes. And my journalist nephew, Sean Stewart, conveyed reassuring words.

Years ago two friends and colleagues sparked my determination to undertake this project: Dan Bausch and Jeffrey Williams, longtime fellow members of the American Society of Tropical Medicine and Hygiene.

Other close friends who provided valuable input and encouragement include Claire Panosian Dunavan and her husband Patrick Dunavan, Eddie Hedaya, Don Wukasch, John Vawter, Dan Myers, Jack Ratliff, and Eddy Rogers. Miles Barne and Lloyd Davidson helped insure the factual accuracy of details in chapters involving them. Professor Barry Hewlett and Professor Vishvajit provided welcomed assurance about the contents of Chapter 6. Ken Burenga used his computer skills to merge several pairs of photographs into single images. Jane Harrigan offered advice and helpful suggestions.

When I first began writing, a talented "Writer's Group" met monthly at the Blanco Library in our nearby town, and these creative ladies welcomed me to join them. The monthly "deadline" thus created a tangible goal, and the group's input contributed to my early progress. Another community organization, the 80 year old Blanco Woman's Club, invited me to a meeting to describe my ongoing writing efforts. The interest and enthusiasm of the members was invigorating for me, and my *promise* to them to finish the project helped hold my feet to the fire.

Finally, I express deep appreciation to Herb Ford, the scholarly director of the Pitcairn Islands Study Center at Pacific Union College, California, for reviewing the Pitcairn materials and for graciously hosting Nancy and me at his Center.

I apologize to anyone I have overlooked.

Made in the USA
Columbia, SC
06 October 2018